Weaving a Way Home

WEAVING A WAY HOME

A Personal Journey
Exploring Place and Story

LESLIE VAN GELDER

THE UNIVERSITY OF MICHIGAN PRESS

ANN ARBOR

Copyright © by the University of Michigan 2008
All rights reserved
Published in the United States of America by
The University of Michigan Press
Manufactured in the United States of America

2011   2010   2009   2008   4   3   2   1

A CIP catalog record for this book is available from the British Library.

Library of Congress Cataloging-in-Publication Data

Van Gelder, Leslie, 1969–
    Weaving a way home : a personal journey exploring place and story /
Leslie Van Gelder.
        p. cm.
    Includes bibliographical references and index.
    ISBN-13: 978-0-472-11642-3 (cloth : alk. paper)
    ISBN-10: 0-472-11642-8 (cloth : alk. paper)
        1. Human beings—Effect of environment on.   2. Place attachment.
    3. Geographical perception.   4. Environmental psychology.   I. Title.

GF51.V3484   2008
304.2'3—dc22   2007037352

For my father, who planted the seeds

And KJ, who always brings the sun

Acknowledgments

This book represents a ten-year journey into the continually challenging and fascinating subjects of place and story. My explorations into the subject have had wonderful company. Some companions helped me articulate my initial ideas, others challenged me to explore more deeply, and others kindly read this and earlier drafts to help me to hone my writing. I am grateful to Sally Palmer Thomason, Steve Misa, and Lloyd Aspinwall for being the best combination of supportive friends and critical readers. Patrick Curry, John Elder, Joe and Helen Meeker, Ron and Joan Engel, Tim and Mairéad Robinson, and the growing members of the Roundstone Conversation, who have kept the discussion going and always offer a new collection of ideas to consider. Larry Morris and the Quebec Labrador Foundation, who gave me a start in the first place, and who provide real evidence for the belief that work which values place, community, and the natural world can and will continue to inspire and interconnect a geography of hope around the world.

During my years at Union Institute and University, and in those that have followed, I was aided by the wisdom and friendship of Nancy Owens, Fred Taylor, Annie Merrill Ingram, Patricia Monaghan, Mandy Gardner, Harry Anastasiou, Yiannis Laouris, Heid Erdrich, and the students of the Oxford Institute for Science and Spirit. In France, many thanks to Jean Clottes for rich conversation; to the Plassard family for permitting us to continue to research and explore in Rouffignac Cave; and to Frederic Gourselle, Severine Desbordes, Marie Paul Abadie, and Nicholas Ferrar (of Gargas Cave) for sitting with us in the glow of a flashlight helping us to answer difficult questions left behind twenty thousand years ago.

I would also like to extend my thanks to the members of my family who supported this project, most especially Gordon Van Gelder, Caren Thomas, Rich Behar, Pam and Jim Sharpe, Miriam Sharpe, Nick Cutfield, Kiri Petersen, Nick Rowe, and Sharleen Maddox, as well as those friends who are family, Evelyn Stark, Mike and Vicky Galow, and Carrie and Ted Opderbeck.

Above all, I am grateful to my husband, Kevin Sharpe, who is the most inspiring person I have ever known, and the best companion I could have imagined for exploring the world. It is a result of his constant belief in me that this book exists, as he has encouraged me every step of the way.

Contents

ONE

Storied Landscapes

When the uniqueness of a place sings to us like a
melody, then we will know, at last, what it means to be at
home.

PAUL GRUCHOW, *The Necessity of Empty Places*

The first people would have had to crawl. To enter the caves
at Les Combarelles, Chauvet, or Lascaux, the first people
would have had to crouch or crawl down long uneven passageways deep into the heart of limestone hills. Crawling on
hands and knees makes it impossible to hold torches in
hand, so the first part of the journey may have been lit only
by a small flickering oil lamp pushed ahead.

A half mile into Rouffignac cave, a place where people
would have scrambled over uneven bear pits to enter, the
first markings appear. In a room known colloquially as "the
serpent room," the red clay walls and ceiling reveal long
white lines, undulating flutings in a cacophony of color and
motion. Deeper in the cave, in a chamber named for the archaeologist André Leroi Gouhran, single-digit finger markings appear on the ceiling in sets of seven. Looking closer,
the pressure of fingerprints left by people at least twelve
thousand years ago linger. Deeper still, at the heart of the
cave where a confluence of black outlined drawings of mammoths, horses, and ibex appear, different finger markings
cut through the soft moonmilch of the cave walls. Long
straight lines made in finger sets of threes and fours stretch

across twenty feet of wall, unique fingers still visible in oblique light.

Ask the guides in any of the caves in which these finger markings appear, and they reply that the lines are *enigmatique*. *Enigmatique* seems to be French for "hell if we know" because, while mammoth heads and shoulders are distinguishable and easy to point out, the collections of line markings that appear on the walls and ceilings of prehistoric caves do not come with an easy guide for translation. Some cave art experts have called these marks "serpents," "macaroni," or "parasite lines." Recently one of the great cave experts of the age referred to them as "the intestines of the underworld."

My husband and research partner, Kevin Sharpe, has been studying these markings for thirty years in both Australia and France. He and I believe that many of these finger markings are not doodles but are instead purposeful and structured, evidence of Paleolithic people's propensity toward writing and storytelling.

No one crawls through caves in near total darkness to doodle on the walls, especially not the people who were gifted artists such as the Paleolithic people who painted the caves in Rouffignac, Lascaux, and Chauvet. These people explored these spaces with confidence. They left behind evidence of their stories.

I stand in bear pits, twelve-foot concave hollows left by the cave bears that wintered in Rouffignac over a period of two hundred thousand years, and try to read the finger markings to Kevin. "A right hand of three fingers crossing over a stream of two fingers ending with a small set of bear scratches." He records our data on digital images as I try to describe the finger markings. We do not know what they say, and yet they speak to us. Sometimes I hold my hands up in shadow across theirs. My fingers are no wider than some, and much larger than others. Women? Children? In the spots where they first pressed their fingers into the soft wall of the cave, I see the pressure of their fingertips. Twenty thousand years from now will our stories be any easier to read? Will our libraries be revered as holy places because they once held our culture's stories? Will archaeologists see our graffiti in abandoned subway stations and marvel that we were willing to build underground mobile shrines for our art?

I follow the footsteps of our predecessors in Rouffignac to bring the threads of their story back to life. Writer Tim Robinson once described his belief that most nature writers are concerned with questions of loss. Loss of stories, loss of landscapes, loss of something we can't even know, but we seem to know we have lost something. With no language to speak their stories, how can I know these ancient peoples' relationship to the cave? I can see with my eyes, but without the stories that accompany the images of lines, or mammoths, they are only marks. With the breath of story, I would be able to see them as they were meant to be. Alive.

At times, trying to decipher the lines, I think of the difference between seeing musical notes on paper and hearing music played. Standing in Lincoln Center or Carnegie Hall with sheet music in hand, I imagine I would feel the same question we face in Rouffignac. We know the place is important for its resonance. We have the notes; we just can't hear the song.

This book is an inquiry into the complex relationship among people, place, and story. As human beings capable of creating sound in spoken language, our species has evolved into story makers. Some researchers believe that *Homo sapiens* may have been the first hominid to have the capacity for complex speech and that the ability to speak has been the hallmark of our survival. While anthropologists have bandied about the terms *Homo faber* and *Homo ludens* as other names for our species, I suspect that *Homo narrans*—the storying species—is the most accurate of all.

Our brains organize around story, making sense of the wash of information we take in with each moment by crafting it selectively into narratives that help us to explain experience and time. Beneath our capacity to story is our deeper comprehension of the concept of place. We are always somewhere, and it is through place that we are able to root our sense of story and our sense of self. Our stories make places important to us, and places become the vessels for holding and keeping our stories. By understanding the concept of place we can illuminate the strands of our most deeply held beliefs, our most intimate and beautifully inarticulate relationships with the world.

Contemporary Western culture has discouraged people from maintaining and articulating deep relationships with places. Few of us live in the homes where we were raised. Today we rarely possess ancestral lands that pass from one generation to the next. In the United States, a country founded on a myth of immigrant people, motion has gained value over rootedness. Speed and time have taken a supreme role over place and relationship, leaving a hollow shell of commodification that encourages negation of local knowledge and nonlinear time. But is each strip mall really the same? What if you owned a store there and your stories lived there? Were the natural places that existed before them more story filled than those that exist now? How would we know?

In a culture in which time takes precedence over place, we run the risk of losing places that matter to us. When we lose those places, we lose our touchstones to stories and deep parts of our identities. Through the demise of language, culture, story, and community, we lose a living connection to the world and the ability to understand our place in it. Instead, we live with a nagging, pit-in-the-stomach feeling that it is getting darker and we have somehow lost our way home.

To focus a book on the subject of place comes with great difficulties. Place is a challenging concept, a coyote of a word, because, although it looks obvious and mundane, the more one explores it the more it becomes increasingly difficult to define uniformly. While academic disciplines, from geography to environmental studies, concentrate on aspects of the lived experience of place, little is written about "place" itself as a subject because it is too large, too hard to pin down. The anthropologist Keith Basso, one of the few who does focus specifically on the nature of place, complains:

> In short, anthropologists have paid scant attention to one of the most basic dimensions of human experience—the close companion of heart and mind, often subdued yet potentially overwhelming, that is known as sense of place. Missing from the discipline is a thematized concern for the ways in which citizens of the earth constitute their landscapes and take themselves to be connected to them. Missing is a desire to fathom the various and variable perspectives from which people *know* their landscapes, the self-invested viewpoints from which (to borrow Isak Dinesen's felicitous image) they embrace the countryside and find the embrace returned. Missing is an interest in how men and women dwell.[1]

I come to the study of place and story not so much as an archaeological researcher and educator but as a person with a storied past that stemmed from my parents' unique view of the world. My father, who, among other things, designed the blue whale at the American Museum of Natural History and was curator in the Department of Mammalogy, believed in the necessity of exploratory experience as the bedrock of a deep education. While my classmates knew of geography from textbooks, my father took his family with him on his expeditions in East Africa and across the United States. At four, in Mozambique, my two best friends were a Portuguese hunter's daughter named Anna Karenina and a local Shangaan boy named Daniel. None of us spoke the same language, but, as children always do, we understood each other well. In our camp, called Zinave, on the banks of the Save River, I played in the sand, had brief and frightening encounters with green mambas in the bougainvillea, and learned how to stay silent while we tracked nyalas in the bush. My father told each of us that we were his research assistants. I believed in that responsibility and was at times enlisted to identify specific antelopes, to listen to radio blips emanating from collars, or put my small hands into the wet grass in the stomach of a dead impala to bring a piece of her to light. I believed I had a necessary purpose and an important place in this research time.

From my father I learned that it was possible to love places. Writing to a friend a few years later from a camp along the Savuti Channel in Botswana, he captured that notion of love of place.

> This probably sounds absolutely mad to you, but there is something quite Eden-like about it. You have all the real needs of civilization—shelter, food, transportation—and all of it comfortable, but you are plunked down in the middle of virtually unchanging wilds, where at any time you see thousands of zebra and wildebeest, giraffes and elephants, and the hippos in the river. Overhead there are birds by the thousands, and at times of the year by the millions, so that they look like smoke when they stream from the trees, and their weight actually breaks the branches. Lions and hyenas abound, and the whole basic cycle of life and death takes place before you.[2]

He remembered the curves of hills, could speak of a particular tree and the cast of afternoon light on it, and knew the animals he researched so intimately that they all had names. One, a small brown-eyed female Nyala, shared my name. For one summer afternoon she and I gazed at each other through a thicket of acacia in the backcountry of Mozambique while my father studied her behavior, her markings, and the graceful way in which she could run. I wanted her to be my sister.

My mother was a poet and writer, the director of an art school and a community organizer. If my father was moved by the curve of a hill, my mother was moved by the hands of a sculptor, the low notes of a cello. She wrote books for children, loved the feel of language on her tongue, and understood resonance through music. She could be so moved by a piece of music at a stoplight that she would not move her car because it would interrupt the moment. She made few friends at major intersections, but she hardly cared as the moment mattered more to her than the chorus of horns behind her.

Both of my parents gave me a life rich in experience. Both died far too young.

I moved into my parents' house in New Jersey when I was twenty-six years old. My father had been dead a year, my mother dead for almost nine. While my parents were gifted at sucking the marrow of life, they were lousy at cleaning up after themselves, and the house I inherited contained the flotsam and jetsam of two overly full lives. From my father's expeditions to South America, Africa, and Mexico he brought animal skulls, carved wooden sculptures, and pottery that we, as children, quickly reduced to shards. Since my mother loved musical instruments, my father found her Burmese harps, lutes made of turtle shells, and inlaid sitars. These were hung on the walls amid masks, carvings, and heavy oil paintings. The concept of "white space" had no place in our home.

Maybe because they were both raised during the Depression, they came to believe that everything had a value. And while they acquired art, musical instruments, and animal skulls with great interest, they also acquired the daily accoutrements of suburban life so that the floor and cabinet space was as full as the walls. Even though they never used the bottles, cans, newspapers, pumpkin pie mix, and index cards they amassed over time, nothing was ever discarded.

At some stage, their collecting grew beyond the confines of value and they could not throw anything out. Unused but still slightly useful items first were piled in the garage and then overflowed throughout the house. Books, magazines, half-finished experiments, pickles canned twenty years before—still waiting to be eaten—mingled with East African art and the curving horns of kudu skulls. Like a poorly run engine, everything went into the house and nothing came out.

The house developed a life of its own, becoming an ecosystem with all of the niches, structures, diversity, and wildlife found in wilder spaces. The seeds of imagination lived there, too, as every item was infused with a storied identity. It became a landscape so filled with stories that they created a cacophony in their competition to be heard.

In the autumn of 1986, my mother took her own life, filling the garage with a carbon monoxide that took her breath away. She died in that house, adding her life's ending to the stories that had written themselves into the walls.

Two weeks after my mother's death, as my father stood paralyzed, trying to pick up the shards of his own life, the museum packed up his enormous office, lab, and collections and brought them to our home. They filled the entire bottom floor of a split-level house. He never once opened a box. Eight years later he died, leaving a forty-five-year career as a mammalogist in a mountain of cardboard boxes.

To live in that house required a reclamation of space.

I moved into the house a year after my father died, having been hired to teach by the sister school of my old alma mater. Living in my parents' house, teaching in the school district I had once attended, I began to experience the multiple layers of time accruing in one place. At any moment in the house, I could travel back in time to various points in childhood, adolescence, or adulthood. Digging in the garden, a space that had ever been my domain, I realized how rooted I was to the concept of place. A family that prided itself on being travelers had a home that told its story, and I, in living in that space, became a keeper of its stories.

The questions of story and place arose through my process of clearing spaces. In sorting each of my father's boxes, I was left with the questions "Why did he keep this?" "Why does this matter?" Sometimes it was an old coffee can filled with obsidian and stones. Sometimes there were seeds wrapped in yellowed tissue. Sometimes newspaper clippings and articles, other times terrible art projects my brothers made in Boy Scouts. Ping pong balls dunked in Technicolor glaze. Tiny calipers for measuring skunk skulls. Cups made of fetal elephant feet. Love letters from his first wife. In each box, I had the artifacts, but without the stories I didn't know where he had found the stones or why he had wanted to bring them back to New York. If I didn't know that I had made the "Chip off the Old Block" project in fourth grade, I would barely have recognized my sad attempt at Father's Day homemade gift art. Were these artifacts any different than the lines on the walls in the cave in France? Without the story behind them would anyone know their significance?

My first husband had no history with the house. To him,

it was filled with junk. Stuff. Papers, books, pictures, meat thermometers, desiccated fish tanks. In my father's landscape, everything had purpose, and when he was alive everything had story. If everything had story, nothing could be discarded or thrown out because all things had a living identity. To him, it all mattered. For me, I had to decide which stories would continue on, tied to their triggers.

In sorting the boxes, I, too, became conscious of my own practice of bringing home touchstones from places that resonate with me. I worry that my stepdaughters will be left with the same set of questions that I faced in looking into my father's tins of stones. Will they know the smooth white stone came from a midnight swim in the Mediterranean in Cyprus or the green stones from my first morning in New Zealand? Will they know the smooth river stones from the inlet in Sweden where their father and I spent the afternoon watching boats go by or the one from the wild beach at Ribadesella that glinted in the morning light just two days after we learned of Kevin's cancer? Will they leave my stones at the cemetery when someday they visit me, as I do with my father's?

While my first husband believed that none of it mattered (and spent time investigating ways to "accidentally" burn down the house), I, on the other hand, was forced to confront my own collection of stories as we took on repairs of walls, floors, and rooms. My father's boxes were an unknown landscape, but in the walls of the house I knew the story. Intimately.

"Who the hell painted over these sockets?" KC complained while scraping paint during our first summer of renovations. I told him the story of the botched attempt to paint the house by my oldest brother and me during one weekend after my mother died. We weren't working on precision, we must have just desperately wanted a clean slate.

"Why are there these greasy marks on the walls?" I heard

as we tackled the living room. I knew it was the spot where our long dead dog, Inja, used to lean into the wall. The oil from her fur must have left a permanent mark on the wall behind the piano. I knew the stories told in peeling paint, chipped tile and rotted wood, the kind of knowledge that comes from living thirty years in one place. How else do you know that a particular floorboard creaks loudest when you somersault over it?

I came to understand the questions posed by storied landscapes through my reclamation of the space. Sorting my parents' possessions left me with the question of what to do with things that had lost their stories. And further confronted with living in the space of my own storied past, I wondered how places accept change. Home, nostalgia, ruin, and displacement were not abstract concepts but my reality. And from my father's fifty thousand slides of East Africa and hundreds of my grandfather's sepia prints of the American West, I was reminded again and again of what it meant to love a place. They had loved places that were not our home. I knew, and know still, that the legacy of people who love wild places, love words, images, and stories, runs in my blood, too. It is not jewelry or artwork that matter to me so much as this worldview. This way of seeing is my most treasured piece of my inheritance.

In the pages that follow is my search into the multichambered landscape of place. As I discovered firsthand, place is the structure of the world, the seat of the self, the root of our relations, and the last frontier of unarticulated intimacy. From understandings of place springs paradox as we love home and travel away, feel rooted and uprooted, alien, and transplanted and all points in between. Not only are places external entities, but we, ourselves, are places— geographical points possessing unique points of view who can feel displaced, replaced, and implacable. Place, by definition, is a trickster, a shape-shifter, because it is so many things at once and yet never all of those things simultaneously. I imagine that if place were a being it would be goat-footed Pan, leaping and changing, challenging us to try to define something that is noun and verb, person and location, from and of. Once upon a time, Pan meant all things. Place, like Pan, is everywhere, and everything is a place. Complexity, comedy, and future breed in such layered, fertile ground, something an old goat like Pan would have liked very much.

Place cannot be defined by systems of hierarchy, only through relationships, like webs or woven baskets. Our understanding of places illuminates our capacity to form relationships, and studying relationships means venturing into the terra incognita of our emotional realms. While the places of the external world can be accessed through maps and a good GPS, our emotional landscapes are best reached through the sharing of stories. How else can we know the mountains, canyons, and wild rivers inside ourselves, and inside others, unless we hear the most extraordinary tales in the world—the stories of how we live our lives?

Story is as complex a term as *place,* perhaps the raven to Place's coyote. In using the term *story* in the pages that follow, I refer to all forms of story, from the conversation with the checkout person at the supermarket this morning, during which you told her how you locked your keys in the car, to the stories that help create understanding of why we exist. They are all story. Many researchers differentiate story into camps of myths, oral narratives, written texts, comedies, tragedies, poetry, and folklore. Stories function like thistle seeds, linking experience to place and one experience to another. What is central is that stories form connections among people, between humans and nonhumans, between jewels of memories we hold. Stories are binding material, what the nature writer Terry Tempest Williams calls "the connective tissue,"[3] serving as "the umbilical cord between past, present, and future."[4]

In the pages that follow, I explore two main ideas. First, that we, as humans, need to recognize our tacit relationships with places because in comprehending our place we understand our capacity for intimacy, relationships, and love. I believe that the solution to environmental issues will not come from a "clean up the mess before it's too late" approach but instead through the exploration of that which we love and what we do to protect those whom we love. To this end, I look at three seminal areas of human-place interaction: wilderness, home, and ruins.

Second, that the way in which we develop and maintain relationships with places is through anthropomorphized relationships with the nonhuman world. When people describe feeling a "sense of place" or "being at home," they are speaking to the feeling of harmony that comes from living *in,* living *with,* and feeling *of* a place. At base, I believe that our unspoken relationship with certain places is one of our deepest expressions of love. That love drives us to act, fundamentally shapes who we are, and offers the greatest potential for shifting the unbalanced relationship humanity has developed with the world around us.

The pages that follow hold my stories and the stories of others who speak to the power and import of place, humans, and story. I hope you find places to reflect for yourself and remember the places that have shaped you and the roots of your own stories. To be human, to be of this place, is to be a member of a community of living beings. Share your stories. I will share mine. Together, we, as members of the storying species, will find a way to weave our way home.

TWO

At the Confluence of Paradoxes
Wilderness . . . Wildland . . . Wild

What's wild is something we haven't yet been able to
define. I'm not sure we can preserve it until we find the
language for it. We have to work very hard and try with
all our energy to find a way to speak what wilderness is.
. . . It's the place of creations.

LINDA HOGAN, INTERVIEW IN DERRICK JENSEN, *Listening to
the Land: Conversations about Nature, Culture, and Eros*

My father studied what many people would refer to as "wild
animals": nyala in Mozambique, spotted skunks in the
Southwest, bats in Baja. Before I was born, two research subjects, a pair of South American coati mundi named Marta
and Jimmy, came to live in his Manhattan bathtub. The coati
mundis followed a tradition of my father's first research subjects, a group of Central Park squirrels that notoriously took
up invited residence in the curtain sashes of his childhood
bedroom, much to my grandmother's chagrin. So it is perhaps not surprising that I was raised in a house where wild
animals (or their skulls and skins) came to live with us—and
the lines between wild and not wild blurred. Kudu horns
made for wonderful impromptu toys. The freezer often contained ice cream, freezer-burned peas, three or four frozen
shrews, and a "study skin" en route to the museum. Bats res-
cued from a neighbor's attic sometimes spent the night in a

tall, clear mason jar beside my father's bed before being released to a new neighborhood in the morning.

Like the animals themselves, my father's seasonal trips to East Africa were commonplace occurrences to us. The familiarity of words such as *expedition* and *the bush* or the knowledge that my father was living among cape buffalo and lions for much of the year gave us a different sense of wilderness than most of my childhood neighbors knew.

Wallace Stegner once wrote that "whatever landscape a child is exposed to early on, that will be the sort of gauze through which he or she will see all the world afterward." Barry Lopez added that this gauze through which we see is "emotional sight, not strictly a physical thing."[1] If they are right, then my vision came through the woven cloth of my father's stories, and those stories helped to shape my future experiences and understanding of wild places.

My father was an inveterate raconteur who crafted his stories with a photographer's sense of character and setting. Of all his camera equipment, he loved his 800mm zoom lens best. To hear his stories was often like feeling the movement of that lens, focusing its sight on each character, each place, pausing briefly as it came into relief. His research trips provided endless fodder for stories. After two months in the bush, he returned home as full of stories as his suitcase was of gifts.

Stories came in three varieties. First, the wilderness stories: these were almost always an outsider tale of the person who refused to engage in local adaptation. These were followed by the code of wildness stories, which recounted the exchange of gifts. And finally came the last variety, wildlands stories, ones in which he shared individual moments of wonder. He did not designate his stories this way; to him, they were all merely conversation about where he had been.

The first variety appeared within minutes of his arrival, just after the hugs and dog petting but before unpacking. While his suitcase sat in the front hall, it looking relieved to no longer be in transit, my father poured himself a cup of black coffee and tried to reacclimate himself to the chaos of the house. My mother had little interest in animals or his research. The only species she studied was humans, and her litany of "research" questions solicited the first stories. "Who did you meet? Where were they from? What were they like? What did they do?"

As he settled into his coffee, we were treated to a kind of storytelling I now classify as "idiot tales." These stories, always similar in theme but different in each iteration, were of people who had no respect for the place they were in or its way-of-being. Traveling to hunting camps, or parks frequented by safari groups, offered my father great scope for a cast of characters who always failed at the simple act of respectful adaptation. Thus we heard the stories of Leonard who spoke with enormous bravado to unsuspecting tourists of his encounters with cheetahs (when the other members of his research team were away in the field) and when they were in camp spent his time popping Valium in an attempt to calm his overwhelming fear of being mauled to death by the animals he feared most—cheetahs.

Then there was the tale of the Frenchman who *was* carried off by a lion one night when he mistakenly assumed he was camped in the Vezere Valley instead of the Okavango Delta. In my father's retelling, the lion picked the man up by the rear end, took one bite, didn't like the packaging or the flavor, and spit him out. In my childhood imagination, I pictured the lion biting into a Twinkie still in its plastic sleeve. Of course, the Twinkie had a head and spoke French. Worse was that this story arrived at the same time that Wrigley was marketing a gum with a campaign that focused on its ability to "go squirt when you bite it."

These two were part of a long-term litany of stories that included people who put their children on the backs of bison, those who ate tuna fish in backcountry camps in grizzly country, and those who didn't know to turn their shoes over at night in the desert and woke to find angry scorpions pacing like disgruntled tenants in their shoes. The worst, though, were those people who insulted the guides and local people. Of these people we heard hundreds of tales, and, though not as dramatic as those about lions and cheetahs, those stories of hubris, humiliation, and arrogance are still the ones that haunt me today.

Idiot tales defied the literary traditions of wilderness stories. They were not Hemingway, Burroughs, or London stories, although those authors were among my father's favorites. No old men were pitting themselves against a marlin while fantasizing about Joe DiMaggio. Nor were people swinging from the trees in scanty loincloths much to the pleasure of their buxom wives and companions named Cheetah. No one was home to answer the call of the wild.

Idiot tales didn't make the wild a fearful unknown but rather the wild the source of a deep sense of respect. It had ways-of-being to which we had to adapt or we were likely to become a bad case of lion leftovers. In idiot tales, specific people caused the problems. Humans in general did not cause the problems—*disrespectful* ones did. In my father's telling, they always paid a price, but sadly, so, too, did their victims.

The second round of stories began as the coffee ran dry. Hauling his suitcase into the bedroom, my father unzipped its cover and searched for treasures buried among the boxer shorts. The act of pulling out the Turkana necklace for my mother, a set of twisted brass and copper bracelets for me, or a Masai shield for my brothers triggered the invocation of the gift exchange. These were stories of gifts we sent and gifts we received.

In the twenty years my father spent traveling back and forth to Africa, he never once came home empty-handed. From these gifts we knew what it felt like for him to be away from his family for months at a time. He could not speak openly of his missing our school plays or birthdays, but the gifts were tokens of a wordless sense that, even though we were not with him in Africa, he carried us in his heart. When he was home with us, he carried the places he loved in his heart. Love was expressed in gifts. This love, deep, with tangled roots, defined the edges of my father's wild. Emotions, passions, love. A place he feared and one that drew him in deep.

Once the luggage had been emptied and the laundry banished to the hinterland, we waited for the last set of stories. These were my father's final gifts, and it was usually days before they arrived. The evocateurs of his wildlands stories, these were the gifts he gave to himself. To me, they are the most profound because he was willing to share them with me. They came in a deluge of yellow cardboard slide boxes from the Kodak plant in Fairlawn, New Jersey.

Many people have coffee tables and accent lamps in their living rooms. We had slide projectors, round black carousels, goatskin drums, and mountains of yellow cardboard Kodak slide boxes. The arrival in the mail of the new block of slides invoked the final, and most powerful, litany of stories: The Moment of Wonder Tales. The slides were not the polished ones he used for lectures, for which one from a thousand might be chosen. Instead, these were the unwritten story of my father's love affair with the animals of East Africa. He didn't make us sit and watch them all as a performance, instead they unfolded through the organic process of sharing a storied passion.

"*Come have a look at this wildebeest . . .*" would come the call, and I would drop what I was doing to sit on the carpet in front of the screen to see the sun glinting in the eye of a wildebeest, elephants emerging from a dust storm in morning light, two flies landing for a moment on a lion whisker. These were the stories of individual wonder, of the savoring of a moment and trying to bring that moment of wordless grace back to the people he loved. With the images in front of us in the darkened living room, we heard the stories of the three-legged lion, learned that lion cubs follow the black tip of their mother's tail through the tall yellow grass of the Serengeti, learned that flamingoes, like flying hydrangeas, get pinker because of the lime in the shellfish they eat. Enormous images of masturbating elephants taught us that our diminutive grandfather, standing at 5'2" and 125 pounds, was the exact same size as the average elephant's penis. Some statistics you just never forget.

The slides served as the backdrop for a long education in the lives of animals. Lacking much plot, these stories had beautiful, memorable characters who invoked a sense of respect and wonder that fed our belief that we needed to adapt to their way-of-being when we were in their home territories. In wilderness my father fell in love, but in the falling of love with those places they ceased to be wild and became more like home. The wilderness became a wildland,[2] and the wildland took root inside him so that he longed for it, just as he longed for us when we were apart.

Paradoxically, my father felt safest in these wildlands. Not safe from physical danger but safe from emotional danger. Like many nature writers, he loved the wild because there he found a refuge where he could heal from the ravages of home.

I did not learn about fear in the wilderness from my father. My mother, who rarely left home, specialized in an emotional terra incognita made of shifting sands and unrecognizable patterns. The concept of wilderness was not a frightening one found in nature to me but was embedded in my understanding of home. Wilderness became my home, and my mother, through the mental illness that plagued the last five years of her life, became the keeper of an unfathomable wild.

I realize that language gets tangled here. *Wilderness, wildness,* and *wildland* are three terms I have used refer to similar places, but they mean very different things. If living with kudus and lions in a tent on the Savuti is not wilderness but living in New Jersey at home with a woman suffering from paranoid delusions and depression is, then what do these terms mean? How can wilderness be home and home be wild? Both are places of heart and head, emotion and experience, story and silence. They find their expression in the continuous dance between feelings of fear, love, and safety in an ever-changing littoral zone between self and other. In wilderness and home are the sinews of paradox, forcing us to swallow our own beginnings and endings, teaching us how to die so that we can live. These are stories of three-legged lions. These are stories of being afraid to open the front door, not knowing what you'll find there when you do.

The world moves in a continuous flow of creation, change, and re-creation. This is the heart of a place-oriented worldview and something my father knew well. His moment of wonder tales and the times when we accompanied him into the bush allowed us to know the cycle of change through the viscera of feeling the thin line between life and death.

In the summer of 1973, we lived in a hunting camp

called Zinave at the edge of the Save River in Mozambique. One afternoon, while out in the bush, we came upon a newly killed impala. My father wanted me to understand the relationship between life, death, and change. I also, apparently, hadn't really understood the concept of an herbivore. We had been eating impala burgers for weeks and had become rather blasé about where our food came from. Here was his opportunity for the teachable moment.

As he cut open the stomach of the impala, he told me that he needed a very important part of the impala, a part so small that it required a tiny pair of fingers to fetch it. To find that piece of viscera required placing my hands through the wet green grass in the warm stomach of the impala. I plunged my hands in, up to my elbows in grass, leaning into the softness of belly fur and hide, feeling the heat of a newly dead body, the musky smell of dust and blood.

With my arm deep inside, I felt the beauty of the impala. I knew her inner parts. We had connected. I could not describe the feeling of knowing something so deep about the wild. I cannot still.

I have no words, and yet I try perhaps because humans are namers who sculpt the world through language and story. To know the nature of wilderness, a place of no names, we must confront the depth of our selves. One feeds the other, and in wilderness we find our boundaries transgressed because we are continually reaching into our unknowns to find, mold, name, shape, bring forth. We want to name that which by its very nature defies language. This is one of the great paradoxes of the wild.

My warm, dead impala, which later became my dinner, could not speak for herself, and yet I heard her. Even now, more than thirty years later, I grope for words and still find nothing quite right. The vacuum created by that wordlessness did get filled by words in my parents' journals, but theirs is the language and interpretation of the observer not the direct participant.

My mother, a Brooklyn native who had no desire to go to Africa and was only convinced to come along when she realized that Bloomingdale's was having a sale on safariwear that summer, wrote in her journal concerning the impala incident of the certain trauma I must have experienced by putting my hands into the it. Raging at my father's callousness, she swore I would end up a lifelong vegetarian because, no doubt, I had been completely traumatized by touching the dead impala. Her journal is ripe with words such as *scarring, traumatized,* and *permanently damaged.* Not to mention her belief that touching a dead animal is *unhealthy* and *disgusting.* There was no mention of the impala roast recipe she had been trying to hunt down only a day or two before, nor did she connect the impala with her supper later that evening.

My father's journal has a different version. His is filled with the nature of the question of wonder. "What if my children could experience what the word herbivore meant personally?" He wondered if he had made a terrible mistake, but he also wrote that he had no idea which experiences would actually stick with me. Any and all experiences had the same likelihood of having lifelong import. He would only have to invest in having faith that he had done the right thing. He was worried but also noted that I had appeared to pull out the little organ he had asked for not with fear or disgust but with a tremendous sense of pride.

My own interpretation in my shiny red travel journal just has the single word *Impala.* Of course, it is difficult to interpret its import because each day generally only had a few animal names as my journal entries. My mother assumed that those were the animals we saw. I am fairly certain it was a list of the animals we ate.

Wilderness has no one voice. I could not tell of my experience of the wild because I knew no words that would be deep enough or full enough. The wild's great power is in its inability to be captured in our words. Yet, because it cannot speak for itself, it has worn the gauze, or crown of thorns, of each culture that has tried to articulate its own relationship with the natural world. We know our humanness by knowing our wilderness. For some people there has never been a wilderness. But it seems for all people there has always been a wild.

A Brief History of Wilderness-Defining Stories

The *Oxford English Dictionary* offers insight into the confusion over the term *wilderness*. Not only is its etymology questionable, as to whether or not it refers to wild deer or wandering, but in definitions it offers specific locations such as deserts and the open sea, as well as "a tract of solitude and savageness." The two are not the same, and yet the term is used to evoke both.

Like *place* and *story, wilderness* falls into a wide range of definitions. The act of defining *wilderness,* unlike many other words, has political and physical implications because the definition of *wilderness* in the United States, as created by the 1964 Wilderness Act, directly impacts how lands are designated and used. While our own personal definitions, like those of *liberty, freedom,* and *justice,* may vary widely from the government-sanctioned definitions, forests, snakes, deserts, and rivers are subject to the following federal definition of wilderness.

A wilderness, in contrast with those areas where man and his own works dominate the landscape, is hereby recognized as an area where the earth and its community of life are untrammeled by man, where man himself is a visitor who does not remain. An area of wilderness is further defined to mean in this Act an area of undeveloped Federal land retaining its primeval character and influence, without permanent improvements or human habitation, which is protected and managed so as to preserve its natural conditions and which (1) generally appears to have been affected primarily by the forces of nature, with the imprint of man's work substantially unnoticeable; (2) has outstanding opportunities for solitude or a primitive and unconfined type of recreation; (3) has at least five thousand acres of land or is of sufficient size as to make practicable its preservation and use in an unimpaired condition; and (4) may also contain ecological, geological, or other features of scientific, educational, scenic, or historical value.[3]

Like the *Oxford English Dictionary* definition, the federal definition speaks to both solitude and savageness. *Wilderness* here is defined by its humanness. It is *not* a place for humans. It is untrammeled and free of human habitation. One may approach in solitude but cannot stay nor take anything but knowledge away. The intent of the Wilderness Act is to deliberately preserve wildness from humanness.

Why preserve? Is it the preservation of animals in formaldehyde or the keeping of butterflies under glass, or is it a belief that humans are the spoilers of something that needs protection from us for us? Perhaps all of those things. Attempts to define the many historical interpretations of wilderness have been undertaken by a number of competent scholars such as Max Oelschlaeger in *The Idea of Wilderness* and Clarence Glacken in *Traces on the Rhodian Shore.* These works give comprehensive explications of the idea of wilderness in European culture from the Paleolithic to the present. Oelschlaeger, along with many others, perhaps championed most by Paul Shepard and Calvin Luther Martin, believe that Paleolithic peoples in the past and more recent hunter-gatherer cultures did not have a sense of wilderness because there is no distinction of otherness inherent in their relationship with the environment. While I believe the descriptions of Paleolithic people are often largely speculative and based on theoretical conjecture, work with hunter-gatherer cultures in the last two centuries supports this notion. For them, there is no wilderness because they do not perceive a boundary between self and world. Instead, hunter-gatherers recognize a similar internal spirit in all beings. They do not see otherness in nonhumans; they see in them a sameness in spirit and a difference in form.

Shepard and others believe that the shift between the concept of feeling *of* a place and feeling *in* a place began in the Neolithic period with the rise of agriculture. Agricultural societies led to the cultivation of species, a rise in human populations, and a separation between a manipulated environment (which came to represent *home*) and an unmanipulated environment (which in time came to be known as *wilderness*). Home was a known place that could become predictable and malleable, yet its creation led to a dualistic "home/away" view of the world as well as a number of other social issues involving dichotomies. Their divinities, entities who lived *in* the sky, became more dominant than those genies who lived *of* the same local environment. Martin writes:

> The real issue, however, is hoarding: the stockpiling of the edible portions of slain plant and animal beings. This is the heresy—a heresy that would be compounded by the Neolithic commitment to artificially producing these plants and animals, stripping them of whatever remained of their spiritual volition, their permission—the gift—in

the process. In the case of plants, it was probably more a matter of repositioning their spiritual value, which I prefer to call their grace, into the theater of the sky gods of weather and time, the keepers of agriculture.[4]

Martin also suggests that the overriding notion of fear is the true root of the agricultural society. Fear of starvation, fear of death, fear of others come out of a culture that supports the notion of a spiritual difference between humans and other beings. The Neolithic shifted spirituality largely to the realm of humans. Even today debates rage as to whether or not dogs have souls. This act, the locating of spirit only in the human, gave birth to the concept of wilderness and its most defining characteristic: our fear of it. We fear it because it is other-than-human and beyond our control. We cannot tame or break its spirit. We do not speak its language.

David Abram in *The Spell of the Sensuous* adds another piece to the historical argument of the movement from being *of* an environment to being *in* it by suggesting that the development of nonrepresentational alphabets irrevocably severed the distinction between living of a place and living in a place. In alphabets such as our own, the language system does not require a tie to a particular environment at all as our letters are not species- or location-specific pictograms. Our place names are often referential only to the names of other people instead of descriptions of the place itself. According to Abram, when written language became fully portable, able to be separated from place and time, the world shifted fundamentally. Nonrepresentational language deepened the rift between knowledge acquired firsthand through sensuous, vocalized experience, reciprocal exchange, and knowledge acquired secondhand through reading. From this comes the contemporary joke about the academic who has tremendous "book knowledge" but little "practical knowledge" of how to survive in the world. In Abram's analysis, the sensuous experience of being *of* the world has become less important than being able to master descriptions of being *in* the world. That split has only intensified over the last six thousand years. He reminds us, "In the Hebrew Genesis, the animals do not speak their own names to Adam; rather they are *given* their names by this first man. Language, for the Hebrews, was becoming a purely *human* gift, a human power."[5]

The notion of the impact of disconnection through language plagues many who are looking for the link between the place-oriented cultures of the past and the time-oriented culture of the present. Jack Turner suggests in *The Abstract Wild* that our culture has worked hard to turn wildernesses, which are real places, into linguistic abstractions that are easily discarded. Once wilderness becomes abstract, it is a simple thing to dismiss; because it is not tangible or relational, it is merely a word. The hollowing out of wilderness through misuse of the word, through simplification in the media, through the manipulation of word and place in the hands of politicians, has left a culture with a strong disconnect between the sensuous experience of wilderness and cartoonish ideas. Ronald Reagan's infamous attempt to block the expansion of Redwoods National Park in 1966 typifies this. "A tree's a tree," he told reporters, "How many more do you need to look at?" Turner's notion is that abstraction makes annihilation easy because the lived relationship has been severed completely. If we don't personally know a tree, why would we mourn its death? His solution is that we must "become so intimate with wild animals, with plants and places, that we answer their destruction from the gut. Like when we discover the landlady strangling our cat."[6] At their root, Turner and Abram both recognize that wilderness has everything to do with language: wilderness's greatest power and its most wicked threats lie buried deep in its and our inherent worldlessness.

With my four-year-old arm extended all the way inside a newly dead impala, I experienced worldlessness. Did I know enough to thank the impala for having given her life so that we could eat? Probably not. Did I think she was somehow defiled because she was dead? No. I had a relationship with her because she touched me in the moments when I explored the insides of her. I could never say what that relationship was because it was something felt but not said. There are no words for the sinews of relationships, but that does not mean that because we cannot name their pull they are not as real.

Historically, the wordlessness of the wild has been its most enduring paradox. Like a clear pool it reflects back to us the story we tell of ourselves. If our society is one based in fear, it fears the wild. If it is one based in the embrace of complexity, we see paradox, humor, and multiplicity. If we fear others, we see others; if we seek resonance, we find harmony.

Our contemporary culture is a direct descendent of agricultural societies. Looking into their mirrored pool, the great story of fear and love, the myth of the pastoral reflects back at us. Is the pastoral a story of wilderness? No. It is the story of the fear of wilderness. That fear, told through the mythical language of the pastoral, is still alive and well today. Even greater than the pastoral, though, is the narrative tradition that spawned it, the one that continues to wield the blade with which much of contemporary culture shapes its perceptions of wilderness—the tradition of the tragedy— our most enduring and suicidal cultural story.

The Penetration of the Virgin: Tragic, Wild, and Ruined

> The "tragic rhythm of action". . . is the rhythm of man's
> life at its highest powers in the limits of his unique,
> death-bound career. Tragedy is the image of Fate, as
> comedy is of Fortune. . . . Tragedy is a fulfillment, and its
> form therefore is closed, final and passional.
>
> SUSANNE LANGER, *Feeling and Form*

High school English students learn that by the end of a tragedy poor Benvolio or good Horatio will be the only one left to standing as he is surrounded by a mound of fresh corpses. With great fortitude he will try to explain to some stable authority figure why his friends are filled with happy daggers, sweet poisons, and s'wounds. Does everyone in a tragedy, except for the stalwart good friend character who didn't take the big risks, need to die? Must Jocasta hang to know that Oedipus has been blind? Surely some therapy could cure that familial problem instead.

Unfortunately, in tragedy, the main character needs to die (or be horribly maimed in Oedipus's case) because tragedies are unidirectional. The main character believes himself or herself to be fated, damned, or beyond the law. That belief forces our Hamlet, Romeo, or Antigone to follow one specific path that leads inevitably to death. By avoiding the exploration of alternatives, death arrives right on time. Distinctions, in a tragic world, are largely based in either/or, black or white, overly simplified situations that are designed to amplify the character's tragic choices and exaggerated character flaws. As the audience, we are drawn through the narrative by knowing from the beginning the basic outcome of the story, yet we strain against it, watching the characters make the poor choices that lead to their dramatic deaths. Through theater, we feel a sense of catharsis for having been pulled through our collection of emotions during the short span of the play. In real life, however, there is no catharsis. Instead, we experience a culture and media following the same language, same scripts, and same modes of thought as Oedipus, who could not see in front of him what was so plain. Are we, too, so blind? If we know those tragic scripts end in death, why does our culture follow them? Who will be our Benvolios? To whom will they live to tell our tales of ancient woe?

Although the tragedy is old, modern writing continues to perpetuate this same dramatic form. One need not read Sophocles or Shakespeare to feel the tug of the tragic. Pick up Jon Krakauer's *Into the Wild* and on the cover of the book itself, printed in black and white, is the sad outcome of Chris McCandless's life. Inside, a hauntingly tragic photo of the young man leaning up against the bus that would become his tomb greets the reader. From the first page we know his fate, and yet we read on, wanting to know why. What was his fatal flaw? What happens to a man who thought he could live off the land? Why does he die in the end? Had he lived, we would not know his name and he would have walked quietly out of the wild having learned new lessons for his own life story. In his death, however, he becomes a central character in yet another retelling of ancient tragedy. The man who pits himself against wild nature is different from the man who goes off to live a simpler, country life. One dies, the other, as typified by Christopher Marlowe's fictional Passionate Shepherd, lives the bucolic simple life with all of the charms of the Neolithic's third space: the pastoral. The tragic tradition is fed by simplification and dualism that makes for easy commodification, consumption, and fantasy applied to all that is not human.

The myth of the pastoral arises from the Neolithic dichotomy between safe, inhabited, peopled spaces and the fear of the otherness of the wild. That is the reflection in the calm pool. Fear of otherness. Because a tragic worldview centers on the creation of dichotomies, tragedy actively opposes complexity, the very source of life, and the defining feature of the wild. Instead, the city, the country, and the wild form a triad. The pastoral, a known landscape of the country, has all of the favored qualities of the experience of being "in nature" but one conceived and tamed by humans possessing none of the fear-based components of being "in the wild." The pastoral allows for the mixture of the human and the animal world. Animals are domesticated, stewarded, and shepherded by their human helpers. The pastoral becomes the landscape of romantic desire—one steeped in the perfume of cottage garden flowers and shaped by the selective, simplistic language of nostalgia.

Complexity makes either/or decision making impossible. Absolutes cannot exist in a worldview of diversity; instead, knowledge is gained through the consideration of multiple points of view and the exchange of stories. Joseph Meeker in *The Comedy of Survival* and Susanne Langer in *Feeling and Form* suggest that comedy is the storied realm of complexity and survival. No stark linear trajectories heading to dead ends, no monoliths, no single amassing of power exists in comedy. Characters live, learn, and love. "Comedy is rarely found in places where power is concentrated," writes Meeker, "but is a daily staple among the powerless. It is the basic strategy for the play of children, and for play of all kinds. Comedy is the way animals solve most of their problems, and it is the message carried by bird song."[7]

Tragedies concentrate power in the hands of the few and create a world easy to consume in small chunks. Interwoven relationships are reduced from layered webs to linear dyads that lead to the all-or-nothing views of the world. In a tragic world based in human characters, the voiceless wilderness cannot speak for itself and is thus incapable of removing its dualistic mantles of purity and desolation. Heaven and Hell. The Virgin and the Whore. Tamed and Wild.

Tragedy speaks to the worst of all human dichotomies: the belief that once virgin purity is entered by humans it is ruined. Virginity, which, like tragedy, promotes the lifeless image of the end of humanity as the distinction of perfection, is saved only for those who do not procreate—a strange form of preservation only the Shakers, perhaps, would have understood. Without sex, there is no life. To make sexuality unpure, a source of shame, is to violate the deepest principle of life. Wilderness, when associated with sexuality in a culture bent on worshipping virginity, leads only to hatred. First hatred of the wilderness for being pure, then hatred of humanity for having "spoiled" it. The wilderness becomes the silent keeper of our shame. And the only way to rid ourselves of this shame is to remove its source.

Turner worries that the language describing wilderness has become abstract and commodifiable. Perhaps he needs also to worry that English has no word for the opposite of *virgin. Fecund. Entered. Violated. Used. Sexually Active. Erotic. Fallen.* Perhaps our culture's difficulties in developing a more complex relationship with wilderness comes from the fact that we never found adequate words to replace *Virgin Wild* once that place has been entered. We lack language for the expression of life force, becoming terminally tongue-tied in the face of the processes of living and dying.

Linda Hogan suggests that the root of many of our culture's difficulties stem from this inarticulateness. "English seems to be a language that has more to do with economics than emotion," she writes, "If we have a language that can't even express all of our human emotions, how can we expect to be able to talk about wilderness?"[8]

The comic tradition speaks the language of the wild

with its embrace of sexuality, humor, and the realm beyond the human. In comedy, fair Titania can spend a night in love with a mule, and we laugh at her words of love lavished on the overly articulate Bottom. The tragic tradition, however, must center solely on the human. Its natural by-product is the bucolic pastoral as the tragic worldview creates a desire for a simplified golden age that has all the components of the culture neatly packaged into a fantasy of purity and simplicity. Thus, the pastoral is always a place of fiction and one just out of reach of reality. Instead, we live with its silent evil twins—disappointment and anger—as we are reminded of the possibility of bucolic purity: new spring lambs bouncing on perfectly rounded hills of green grass where we have all the time in the world to contemplate the gently passing clouds. The pastoral is the unspoiled fantasy put on the escalator of time that allows us to forever "look back" on a purer and less complex time. Time becomes the great winner in the myth of the pastoral. The elusive purity, and simplicity, which we are taught to crave, always lives beyond our grasp in the romanticized past while we are left with an unnecessarily complex, ruined, and violated present.

The pastoral is not about the arduous work of tending to animals, tanning hides, or predawn milkings. The pastoral is not about the castration of males, the shoveling of manure, or the destruction of crops in a sudden May hailstorm. Instead, the pastoral is the sunny day when the bluebird sings from the lemonade springs at the big rock candy mountain. The pile of carcasses burning in a meadow, roasted and sacrificed for having the possibility of foot-and-mouth disease, is not pastoral. No Passionate Shepherds write poetry to nymphs near flocks of bleeding sheep.

What is wilderness in an anthropocentric world where the pastoral fantasy of nature is the desire? Wilderness comes to bear the weight of the dichotomous relationship of the pastoral at its most extreme. While the pastoral is a pure place because humans live in harmony with a tamed version of nature, the wild becomes either the apex of purity because it is devoid of humans or the pit of terror because it is untamed by humans. In both cases, the role of humans in the environment becomes what is central to the story. Wilderness ceases to be a complex place of its own being and instead becomes the repository of extremes as each label is placed upon it and a story is built to support it. Abandon hope all ye who enter this dark wood. The dark wood will not tell you its tale.

Recently, a genre of new wild tales has begun to emerge, like the types of stories my father told, replacing the tragic literary model with a comic one and replacing the virgin wild with an erotics of place. These types of stories offer a clearer path for illuminating complexity in the relationships we have with the wild. Terry Tempest Williams writes of an erotics of place; Paul Gruchow speaks to the grace of the wild; Gretel Erlich finds the solace of open spaces. *Erotics. Grace. Solace.* These are terms from our emotional landscapes, and yet they have come to form the central language of a new form of writing, which illuminates the power of our individual personal relationships with wildness, an environment woven into the internal through the outside.

Gary Snyder's *The Practice of the Wild* was one of the first to attempt to speak to the complexity of the concept of wildness. Coming from fields such as anthropology, spirituality, and poetry, Snyder suggested that we need to be conscious of our experience of wildness and need to learn or relearn how to develop active, lived relationships with the world around us. Thus a practice of the wild is about a way of being in the world, a complex, lived, deeply storied relationship. Snyder tells us, "To resolve the dichotomy of the civilized and the wild, we must first resolve to be whole."[9]

Meeker suggests a route to that wholeness through the recognition of the power of the comic model, one in which diversities flourish and play creates the edges of chaos and forges passageways through which novelty can emerge. Meeker reminds us that creation and continual creation are the end goals. Unlike the outcome of the tragic tradition, in the comic, *life*—in all of its unpredictable ways—is the ultimate outcome. The comic is a story we learned from the wild, not a story we imposed on it. These are relational tales, stories like the ones that grew from my father's slides of early morning light piercing the Rift Valley. While the bookstores continue to be filled with Jon Krakauer's *Into the Wild* and various tales of climbers, rafters, and mountain bikers who "pit themselves against nature," an equal number of stories are surfacing of people attempting to articulate their emotional relationships with places they call wild.

All stumble over language. All wish to be eloquent. All are still looking for a language for love. Reading them is to resonate with their struggle to articulate that which gives us solace, a sense of grace, beauty, shared grief, and access to the spiritual dimensions of our lives. These new stories are rooted in what might be some of the oldest human story-telling traditions of all, ones we learned when we didn't know a word for wilderness at all.

THREE

The Intimate Wild

> Wildness matters not because it alone is sacred but be-
> cause it arouses in us the sense of sanctity that makes vis-
> ible the sacredness of everything else in life.
>
> PAUL GRUCHOW, *Boundary Waters: The Grace of the Wild*

The most terrifying wilderness I know is not located in an
unnamed dark forest in a northern country, nor is it in the
tawny yellow grass that hides the eyes of my father's lions in
Mozambique. My deepest wild, the one I always need to explore but the one I have trouble knowing how to enter, and
of which I'm terribly afraid I won't be able to leave, is located just below my heart in the pit of my stomach.

There, in the dark green places inside of me, is an unnamed collection of feelings, a place where faces, and
frozen moments of time, bubble up and catch me unexpectedly in the throat. There is my wild. Moments of wordless grace live there, too—memories of a raven's feather
drifting down from the sky to my feet, memories of my father's eyes closed for the last time beneath a white hospital
sheet, the feeling of a car spinning out of control on the ice
seconds before being hit by a truck twice its size, the mobile
phone ringing as we crossed into Andorra with the news of
my husband's cancer—there, in that place of my own creating, but the one in which I feel I have the least amount of
control, is my wild. I feel it but have trouble naming it. I go
there unwillingly, but to avoid it is to let it grow so dangerous

that I cannot escape from it. Fear lives there. Fear breeds there. Love lives there, too, and I know it because I can only barely begin describe the beauty I feel when I catch the moon shining through the window onto my husband's sleeping face or see late season thistle go to cotton and ride the wind. Known and unknown are rarely separated out because I do not want to venture into a place from which I am not sure I know how to return. My wild is an undifferentiated landscape dominated by untamed feelings. The only person who can tame it, paradoxically, is me.

Each of us possesses a wilderness of our own. We all seem to know the contours of the rivers of our own encounters with life and death. In wilderness, in wildness, we illuminate those relationships. To understand them and recognize their necessary power is, to paraphrase Thoreau, the preservation of the world. Our own survival is continually at risk and depends fully on our wildness. At the confluence of our paradoxes we embrace our wildness because it is the deepest part of who we are. And the place of our own creation is the one without words. We can learn its way-of-being by finding resonance in the worldlessness of the external wild. To know ourselves we must know others. Wilderness is not an entity in itself; it is a relationship bound by our understanding of place.

In the summer of 1985, after spending the previous two summers in East Africa, I got a summer job teaching environmental education in Newfoundland. Although I was plenty familiar with kudus and elands, I knew almost nothing about boreal forests and even less about the easternmost province of Canada. For the next five summers, I worked in various places along the Newfoundland and Labrador coasts. It was there that I had my first personal understanding of the way in which wilderness functions as a system of relationships.

The backcountry of the Lower North Shore of Quebec is some of the least human-inhabited land in North America. Blackflies swarm like biting banshees. A calm, windless day may look beautiful in photographs, but Dante should perhaps have located a ring in the Inferno specifically near a place called Baie de Ha Ha and let those who have committed some horrid act spend their time trying to portage a canoe over untrammeled muskeg while the flies are freed to feast. Perhaps because of the "wild animals" I was encountering I could not appreciate so well that I was experiencing what would best be known as a textbook example of a "wilderness."

The inland country beyond Baie de Ha Ha (which was no laughing matter) was not currently inhabited by humans. People may have passed through in the past, may have "left no trace" while camping there, but by and large it was devoid of contemporary human habitation or signs of direct manipulation. There were no mines, no signs of logging, no hunting, no evidence of people taking from the land. Perhaps in winter there had been traplines through this area in the past, but few people in the local community spoke of still engaging in winter hunting. There may even have been one or two moose that had not heard of humans before. Unfortunately, to the flies I was a rare delicacy.

One of the first characteristics of a wilderness is that it is a place where humans do not dominate. Instead, it is a place where the interrelationships of nonhumans set the tone. That does not absent humans from participation—I was clearly engaging the blackflies—it only suggests that a wilderness is an environment that is *not* being directly manipulated by humans. I had not drained the streams in which the blackflies bred so that it would be more inhabitable for me. All of my attempts to change my own physical environment, namely, by pouring far too much Muskol all over my skin, only left me feeling that strange toxicity that only Muskol can produce, especially when you notice that the clear fluid you poured on your skin has just eaten through your raincoat's sleeve.

What is powerful about a wilderness is that it is unknown and unfamiliar to us. This is a terminology of relationship. I had never been to the backcountry behind Baie de Ha Ha, but I also knew little of inland boreal forests. To Bill Robertson, who ferried me out to my drop-off point at the head of the bay, it would not necessarily have been a wilderness but a wildland. I suggest the use of the term *wildland* for a place that is not dominated by humans but is known to us as a distinction between *wilderness* and *wildland.* While blackflies and red-throated loons might not be able to know the distinction, as the land through which I portaged probably felt very much the same to them, my own set of ways in which I related to a wilderness were different from the way in which I related to a wildland. A wildland has safety in familiarity because there is a set of relationships in place, a wilderness does not.

For me, landing at the edge of Baie de Ha Ha and setting off into the wilderness, I can say that it was a wilderness because it was a place of unknowns, a place of "others" with whom I had not yet developed relationships. I did not know the outline of the glacial hills at the edge of the bay. I did not know which animals were indigenous to the backcountry nor did I know the speed at which I might be able to walk through muskeg with a heavy pack and an ancient canoe. I had a map, a compass, and a wide-open northern sky. In that wilderness, I had more fear than I had ever had walking through Manhattan late at night. This is perhaps not surprising because in wilderness we carry a strong sense of fear, which serves to heighten our sense of the boundary between self and other. We do not have prior knowledge of all patterns of behavior. Unpredictability is unnerving. I could not know if I would find a moose or a wolf beyond the next bend. I did not know in advance that the coastal breeze in Quebec and Labrador keeps many of the flies at bay. Only going inland did I discover firsthand why some explorers had died from blood loss incurred by swarms of blackflies.

I did not know how to read the clouds nor did I know the directions of the wind. My fear was bred in the novelty of my situation. As a person with few great claws or very pointy teeth, I, like the rest of my species, came to rely on my capacity to predict as my adaptive advantage. And when we, as humans, do not have enough knowledge to be able to predict accurately, we respond with our strongest biological response: fear.

In wilderness systems, diversity, adaptation, cycles of creation and re-creation, interdependence, and complexity dominate. We, in our relationships with nonhumans, look for either affirmation of our difference (often through conflict) or resonance in our similarity. I did not feel great resonance with the flies, although their biting had reminded me of some of the ninth-grade students who used to swarm around my desk at the beginning of class, so I felt comfortable affirming our conflicted difference and swatting or squashing as many of them as possible. This feeling was not true of the graceful red-throated loon that swam ahead of me as I paddled across Robertson Lake late in the day. I slowed to watch her, to study the way she moved, and then when I put my paddle to the water I wanted to move as soundlessly as she.

I knew I was still a human and she still a bird, but in the wild we continually test the borders of what is self and what is other to discover the lines of resonance or dissonance. Many cultures refer to this through stories of shape-shifting. While I did not want to necessarily become the loon, in trying to move my paddle through the water in the same way as she moved, I was trying on her way-of-being to see if it fit me. This shape-shifting, simple in my case but often more complex for others, runs with the risk of the potential fear that we may be afraid of losing ourselves in the process. From my own experience, I believe we always find more than we lose.

Wilderness is sensuous. Wilderness is wordless. We can lose ourselves or find ourselves there. Alone but not alone, we find wilderness at the confluence of paradox, riding the curve as inner moves to outer before moving in again.

We need solitude to be open to the experience of shape-shifting. If we are continually narrating to another human, we are using our common language of explication. Our potential to break out beyond the realm of our species is compromised. Freed of human language, in the wild, we discover the simultaneous capacity to be both wordless and the creators of new language.

Wilderness becomes wildland when it becomes known to us. Fear is replaced by recognition, memory, and story. In wildlands we recognize the possibilities of resonance. In wilderness, our fears make us cling to the known. Wildlands still contain the same natural laws of chaos, complexity, death, interrelationship, and adaptation, but we also feel a sense of safety in that we are in an environment where we are known. We have relationships. We know names. The red-throated loon and I know each other, and I know something of the way in which she glides through still water. I know her barking call.

In wildlands, we are able to fall in love. We are still free to be feasted on by blackflies, but in the relationships we have developed with others, and them with us, we have a middle ground, a storied landscape. This is the nature of a wildland. We have no less possibility of being eaten by a lion, or drained of blood by a cloud of blackflies, but we have more knowledge of where the lions, or flies, might be and how they might like us prepared.

This wildland was the home of my father's moment of wonder stories. His wild was one of known characters, known locations, and animals whom he expected to behave in the ways he knew. Lions eat Frenchmen in sleeping bags when left out at night. Green mambas strike at small children with curious hands poking about in the bougainvillea. Nyalas are shy of humans and live deep in the bush. Behavior does not change, but in a wildland one can predict through relational knowledge.

Although I went into the land behind the Baie de Ha Ha seeing it as a wilderness, I left it with the relationship of a wildland. I came to know the way of the wind across the muskeg, know the shape of inland lakes and trees, know the winged creatures and their bites, know the shape of the sky. Although I have not been there in almost twenty years, the place would not be unfamiliar to me now as my relationship there is etched in me.

Perspective and the personal are key here as one person's wilderness is another's wildland. This is why Wallace Stegner wrote in support of the 1964 Wilderness Act, "We simply need that wild country available to us, even if we never do more than drive to its edge and look in. For it can be a means of reassuring ourselves of our sanity as creatures, a part of the geography of hope."[1] We must know our edges, forge complex pathways through the terrain of self-other, recognize ourselves as creators in the world of creation if we are to know who we are and how to live respectfully. I learned my edges in that wilderness and the many others I have known throughout my life. I am still learning as there will always be new wilds, for even the "conquered wildernesses" of the world will still be very wild to me.

Wild Words

The landscape of Quebec and Labrador etched itself into me and shifted from wilderness to wildland in part because of the language I developed there. When Jacques Cartier first sailed down the Labrador coast he called it "The land God gave to Cain" because he could not see any resources worth exploiting and the land looked very much the same. Monotonously the same. He did not look close in, at the sundews and cotton grass, the spread of spruce trees trained by the wind to grow sideways, or the rivers full of silver salmon. From the coastline he saw an undifferentiated tundra. A wasteland.

My arrival on the Labrador Coast on a snowy July day brought the same shock to me. No trees. A world of dull brownness with little growing higher than my knee. I did not know the names of birds, fish, or trees. I had no knowledge of which berries would feed me and which would kill. They all looked the same. Just berries.

In Africa, my father, the mammalogist, did not often take as great an interest in birds. Although he knew most of the larger, flashier birds, when it came to the smaller birds he often referred to them as the LBJs. The acronym had nothing to do with presidents from Texas but instead referred to a category of undifferentiated bird known as "little brown jobbers." Even though they were all different species, without names they became very much the same.

When we venture from a known environment into a place we perceive as wild, our first step comes with the immediate desire for differentiation. Are all the trees the same? Is every berry edible? How do the hills appear here? When I was first let off in the Baie de Ha Ha none of the hills was familiar to me. In my fear of not knowing where I would need to go to be picked up days later, I imprinted the out-line of the hills and the inlet in my memory, where they still live today. In the wild, we locate by the specific, not the general. Through relationship we build knowledge by relying on what we already know, so we can separate out the known from the unknown. John Muir's old adage holds true: "When we try to pick out anything by itself we find it hitched to everything else in the universe."[2]

Our relationship with wilderness triggers the process of differentiation. Differentiation moves through rapid-fire simile relationships. "That flower is *like* the one in Grandma's yard only smaller. . . . That cliff is *as high as* a two-story condo . . ." Familiar faces appear to us, too. "That crag *looks like* Aunt Bertha's nose." In time, the longer we stay, language moves from simile to metaphor. The craggy mountain ceases to be just any mountain and becomes Aunt Bertha's Nose. When it volcanically expels its ash, it is naturally Aunt Bertha's Sneeze. Less dramatically, mountain names such as Sugarloaf or Tetons speak to other explorers' propensity to use metaphorical naming. Metaphorical language creates layered relationships with multiplied meanings and engagement in complexity where a name can simultaneously signify more than one thing, such as funny noses, lump sugar, or enormous breasts, as well as the mountains in front of us. What is most important is that they are connected through us into storied landscapes.

Although we begin with an undifferentiated place when we enter into a wilderness, in a short time, to locate ourselves in that place, we have begun naming. To name things with meaningless language devoid of a specific relationship does not help us because we need to attach our feeling of unknown to that which is known. Our survival depends on it. Aunt Bertha's face follows us and makes the world around us familiar and turns mountains into family.

Naming is intrinsically powerful. To name is to define, to call into being, to differentiate from unknown to known. Edmund Carpenter writes in *Eskimo Realities* of the Inuit approach to language not as a form of labeling the known but as calling forth from formless into form. "Words do not label things already there," he writes, "Words are like the knife of the carver: they free the idea, the thing, from the general formlessness of the outside. As a man speaks, not only his language is in a state of birth, but also the very thing about which he is talking."[3]

Think of a child's first words. You probably don't remember speaking your own, but your parents remember them. They waited to hear you speak them. They waited for *you to name them*. When you named them, they knew that you knew who they were to you. Differentiated individuals. Unique.

Trees will exist whether we call them trees or not, yet we may not be able to know them as trees unless we name them. They will remain trapped in the collective. It is this collective abstraction, a nameless uniformity, against which Turner cautions in *The Abstract Wild*. Once we have named the trees around us, or the mountains and hills, we have engaged them in the same way that we engaged our parents. We have woven a relationship with them because the act of naming has put them inside of our imaginations and brought forth a name that links us. The nature writer Paul Gruchow emphasizes this when he writes, "Imagine a satisfactory love relationship with someone whose name you do not know. I can't. It is perhaps the quintessentially human characteristic that we cannot know or love what we have not named. Names are passwords to our hearts, and it is there, in the end, that we find the room for a whole world."[4]

In my summers camped along coastal Labrador, I spent much of my time watching waves. I was fascinated by the motion of process, one in which water came in, moved back out, the same and yet different again. The process of differentiation functions similarly in that it is a cumulative effort but much more like an ocean wave breaking on a beach. We feel the present moment of the cresting of the wave, but that wave is made of all of our past brought forward into this new present moment. When it returns back to sea again, we have added our new knowledge. The new waves break again and again with more of our past incorporated into the view we hold in our present, and we grow.

This image of the wave is similar to the way in which Confucian art depicts time as a river where we humans are facing downstream. Our future rushes up behind us unseen. We sit on the rocks as our present glides by but are unaware of its continuous presence. Instead, we look on our past as it unfolds in front of us in an ever widening river. In the pools and eddies, rapids and flat waters we have a fluid, connected motion to which we must always assimilate the new into the patterns of the old.[5] Into this image of time flows the places we have been and our memories. In the waters, we look for the anchors of place to help us connect the continually unfurling pieces of experience into narrative stories.

Developing continuous relationships, we internalize the shift from wilderness to wildland as wonder replaces fear at the very crest of the wave. At some stage in the process, we stop looking for difference; instead we want to know how are we similar. Do we know the same feeling of a fern frond unfolding? The delicious moment of a raindrop hanging in all of its fullness from the lip of a wild rose petal? Do we laugh at two bear cubs rolling over each other in afternoon light? Or dip a paddle into a still lake with the same gentle ease as the red-throated loon?

To be in relationship with a place, to know its patterns, its way of being, is the way in which we become a part of the place and the place becomes part of us. Otherness and self blur. Only we know when a place has moved for us from the idea of an unnamed wilderness to the reciprocal intimacy of a wildland, just as only we know when a stranger becomes a friend. For some, the distinction never takes place. For others, perhaps those knowledgeable in many different ways of being, finding an external wilderness may be impossible. If we stayed long enough in any place, a wilderness would turn to wildland and wildland to home.

Few of us have the opportunity to experience the full evolution of relationship from wilderness to home, but we feel pieces of it and become at home in wild places. This act of homing is the very root of the expression "having a sense of place," what Thoreau describes as the essence of the saunterer, the subject of his essay "Walking." He writes, "[S]ome, however, would derive the word from *sans terre* without land or a home, which, therefore, in the good sense, will mean, having no particular home, but equally at home everywhere."[6]

Through internal evolution from wilderness to wildland we learn how to become at home in the world. Although our journeys are our own, they are always about our relationships, and in that way they are forever linking and connecting us to the fabric of other lives. As the Lakota expression "Mitakuye Oyasin" ("All my/our relations") suggests, we continually engage, whether consciously or unconsciously, with all of our relations.

Iteration: Storying the Journey Home

My first understanding of wilderness came from my father's stories. His idiot tales, his gift stories, his moments of wonder stories became the framework through which I first came to understand his relationship with the wild places he encountered. Although his trips were largely without us as companions, he brought Africa home to us in word and picture. Some stories were only told once, in the hours after he returned from the airport in the full rush of the pleasure of being home. Others, the stories that became the stuff of legend, were told over and over again, each time becoming more and more polished and smoothed with the retelling.

Meeker writes that this repeated storytelling, in which the story is crafted into a well-timed, well-told tale, is part of the comic tradition of survival tales. Patrician Monaghan suggests that "Storytelling is based in iteration. It's not a story if it's told once. It's only a story if it's retold."[7] The word *iterate* comes from two sources. In one, it is from the Latin word for repetition, *itero,* meaning "to say again." In the other, it is from the Latin word for a road, or journey, *iter,* from which we get the word *itinerary* today. Perhaps the words are more linked than we think, as the stories from journeys are repeated over and over again, evolving with each new iteration.

A journey into the wilderness allows for the formation of new language as we place ourselves through the process of differentiation. When we return home, we want to share our stories, whether they be of Frenchmen carried off by lions or bloodthirsty squads of blackflies. The question is, how do we take a place that now lives in our imaginations and bring that place to others in a form they can experience without ever having been there themselves?

The story might begin with, "The place where I camped was at the foot of a mountain that looked like Aunt Bertha's nose" or "The lion thought the Frenchman was like a giant Twinkie still in the wrapper." For the most part, there will always be inadequate words for the experience of transformation because we cannot bring the three dimensionality of the experience to life.

Most of the experiences of being in wilderness are not the stuff of story; they are the experience of process. Narrative structures, especially ones in a tragedy-oriented culture, require a setting, rising action, climax, falling action, and conclusion. What if our experiences don't fall into those categories? Reflection, walking, listening but not speaking, being scared or cautious or silly, discovering that you are able to sing the entire Beatles oeuvre loudly and off-key, or swimming naked with frogs and a surly beaver are not the stuff of narrative. Developing complex relationships is rarely the stuff of story because it does not follow a linear narrative pattern. Instead, we find collections of moments in time etched into our inscapes, like a collage of pictures, whispered haiku, or the images captured in the yellow boxes of my father's slides. We do not always have stories of love, or pain, because these are moments in time that we feel but cannot narrate. Love that has grown over decades is often more easily captured in a photo of two gnarled hands twined like ancient ivy than written in a plotted narrative. No one looks for "falling action" in their own lives.

It is hardly surprising that we rely on photographs and artwork as a way to express the primacy of the wordless experience of place. For my father, the slides evoked his real passion for his work. While photos can serve for some people as the modern day equivalent of shooting a buffalo and bringing it home to mount on a wall, for many other people photos serve a more important purpose. They capture a stop-time moment that will never clearly exist in the plottedness of story. We cannot bring someone else to the feeling of a red sun rising over a glassy lake, the feel of spring grass beneath bare feet, the sound of swans beating wings against a river in the moments before a storm. But we feel those things just the same. Stories are our way of reaching out from our own wild to touch another's. We don't want to be alone with the force of our love, the weight of grief, the breathlessness of beauty. We try to share, and in that we find wilderness offering us the gentle hand of the geography of hope.

I hear the rise in my father's voice as he points out the small tear in an elephant's ear. I read his letters that end with, "I want to take you here some day." I catch grains of pollen on the edges of a spring willow with my lens and send them, with no words, to a friend who lives far away because I want to give him that moment of possibility and rebirth when he's forgotten it. I have never forgotten the red-throated loon I knew on Robertson Lake.

When we cannot let our wordlessness speak for itself, or when we are prompted to performance, we fall back on the "survival stories." These are the notorious Hemingway, Faulkner, and London stories that high school textbooks use to delineate clearly the theme of "man versus nature." Other stories in this genre are of the wily Odysseus variety, centering on how we survived our misfortunes. Poor planning makes for a good story, but only after the fact and only when we have survived the situation. The night camped on top of Mount Tammany with no water in freezing winds with an obsessively barking Labrador retriever makes for a good tale only long after the blisters from poorly fitting boots have healed.

One essential aspect of our encounter with wilderness is that we are creators there. Putting our words to wordlessness, we create the manner in which we differentiate. Because we do this through language, we are open to the dangers of abstraction and exploitation. The wild's silence in our language is not tacit approval of its annihilation.

That is why so many are moved, through quiet relationships with places, to become voices for wildernesses they perceive are being violated. To cut a tree I love is to cut me. Where is the border between self and other? We only know our selves by knowing others. If we remove the others, we will never know who we are. How will we resonate?

Although language has a profound influence on shaping our experience of wilderness, the most profoundly moving experiences—the ones in which we encounter fear, grace, generosity, ferocity, and grief—are the ones we don't turn into tales at all. These are our deepest encounters with the wild.

Death, Eros, and the Wild

I first read Barry Lopez's *River Notes* during the November darkness of my second year at college. My mother had taken her own life the fall before, and my grief took on an inarticulate flavor, one in which I found solace in a wordless relationship with the trees and dancing autumn leaves at Nichols Arboretum but could not speak of to the people closest to me. The Huron River knew me, and I it, as I spent hours watching black walnut leaves swirl down its banks into darkness.

Two seasons later, I moved to Labrador for the summer. My pack had room for only one book, and I took *River Notes*, into which I inscribed Robert Penn Warren's poem "The Place" over the dedication page. The words of Lopez's dedication, "for Sandy," emerged as two rocks anchored in the face of this stream of Warren's words.

This
Is the hour of the unbounded loneliness. This

Is the hour of the self's uncertainty
Of self. This is the hour when
Prayer might be a possibility, if it
Were.

The poem ends with the comforting thought that "Nothing astounds the stars / They have long lived. And you are not the first / To come to such a place seeking such difficult knowledge."[8]

When my father died eight years later, during a winter of endless ice storms, I curled up like the aching pines and thawed only through reading Peter Matthiessen's *The Snow Leopard*. I walked for days along the Delaware River, looking for proof that spring follows winter and that even glacial ice eventually melts, leaving warm, smooth valleys.

These writers, gifted in recounting their experiences in the wild, take on the impossible task of giving voice to the wordlessness of grief. Lopez, Warren, and Matthiessen speak to the encounter of coming to natural places alone searching for language, wisdom, a way to make sense of an internal pain. All describe the encounter of the self sacrificing self in the face of nature only to find no wisdom given by the natural world except an understanding of the way in which we are so like all things that live and die. Their writing of resonance carves a stone path from the wordlessness of grief to the development of intimacy to the reclamation of voice and love. With human company too difficult, because it asks us to find words for that for which we have no words, the solace of the natural world offers a level of resonance into which we are always invited and through which we can be healed. These writers understand the power of hope. Theirs is the literature of the wild.

The wild is the heartbeat of existence, the life of the world and the death of us. It is the confluence of paradox that flows like fire through our veins. Wilderness and wildlands are relational locations, places to which we offer our designations though our relationships with them; wildness is a state of being. We don't go to wildness; we feel it. While we go into the wilderness, wildness comes to us, unexpectedly, powerfully, with the force to create life and take it away in the same breath only to make something new from the old again.

Wilderness is an environment of undifferentiatedness where the patterns of behavior are not known to us. Wildlands are familiar but untamed, interconnected and free.

Wildness is the frontier of creativity, the movement of the border zone between self and other, the place where death gives birth and life welcomes death. The place where we must lose ourselves so that we can be found. The home of paradox.

My wild lives in the pounding surf on ancient rocks off the Cantabrian coast of Spain. Hurricane winds beating against maple trees in my yard. The swirl of green against black sky and the sound of tree trunks testing their limits to see at what moment they will move from whole to broken. Kevin's hand on the small of my back when he kisses me.

First times are wild. Openly engaging in the elements of chance with the willingness to not know where it will go. Tapping into the creative force of the world, feeling resonance with others who are there, feeling it deep inside. Taking a risk. Writing the first line of a story without knowing the character's name. Risking the potential to be "out of control" and willing to bathe in heart-stopping beauty or savage pain. Wildness is the state of being open to the unknown. Putting faith in embracing its power to transform life and death. Calling into form and then releasing to formlessness. A sensuousness born in death bringing life. Seeds of chaos. Wordless intimacy.

The experience of wilderness does not need to be voluntary, but we choose wildness. It can be more terrifying than any wilderness. Why would we as rational human beings want to experience wildness? Why would we choose to know death with the same intimacy with which we know the smoothness of our lover's hips?

Wildness is the bubbling spring of change. To engage willingly in the dance of creation and re-creation, often through what appears to us as destruction, only to feel the ripping power of creation again, is putting our fingers on the very source of being—a river of cool liquid fire. To deny wildness is to court finality.

In my parents' deaths I searched for life. I could not find the words for it, but I could know it in quaking aspens, white birches, and red leaves drifting down to the sea. It still catches in my throat. With the red-tailed hawks that fly over their graves I never need speak, but I know what it feels like to soar, to continue on.

Change invites us into the world of chaos. Mythologically, worlds are born of chaos and eventually develop so much complexity that they return to chaos again. That enormous breathing cycle of breaking down, complexifying through continual change, and breaking down into smaller parts again is the rhythmic heartbeat of our planet's way-of-being. When we open ourselves to randomness, to novelty, to chaos, we open ourselves up to the power of creation, to the power of dying, to the spirit of renewal.

Faith grows in the damp green wetness of the wild. A sense of wonder, a sense of resonance, a sense that we are able to retain the edges of ourselves but see our love, our pain, the flux of living and dying in others and to reach out. We embrace the wild to know the dance of creation and destruction not as an abstract but as the marrow of life.

How can we make sense of a place, a state, that is by nature unknown? We cannot. What we can do is to learn the secret of return. The journey. The iteration. Wildness flows, pulling us toward the place where we lose the boundary between self and other.

If we are willing to love selflessly, to sacrifice, to hold the hand of our dying parents, to cup a bird with broken wing in our hands and feel its heart slowly stop, to tuck the petals of a fallen poppy into our hair, we have known the grace of the wild. When we have learned how to be open to living and dying through gestures of compassionate honor, respect, and gifts of weightless love, we have learned the language of the wild. We become another voice to a song, a dance no one sees but us. We offer our last cup of water, for we have indeed found the source of the river of paradox. We can drink of it fully as it will only wash through us again. This is the sacredness of the wild: sacrifice, the open heart, the selfless gesture. We lose the boundary between self and other by choice, and in doing so we find the deep spring of life in that selfsame moment. Compassion. The dance of creation.

In the wilderness we come to know our boundaries, know that from wordlessness springs language, from aloneness springs relationship, and from death springs life. When we lose a sense of wilderness, lose the capacity to create relationship with nonhumans, we develop a language of abstraction, commodification, tragedy, which allows us to take the spirit of life from the world. We will find death but not know the path to life again should we forget to learn the graceful lessons of the wild.

We are places, filled with tempests, calms, memories, and stories. Our journeys through the outer and inner wilds offer us the chance again and again to know who we are and to engage in the wild energy of that which flows through all beings in the world. Life. If we are at home in the world, we must learn to be at home with our wildness, for in it is the truest nature of who we are. Beautiful.

The Intimate Wild

FOUR

Always Leaving, Always Coming Home

We learn our homeland from stories, just as we learn
nearly everything from stories.
    STEPHEN TRIMBLE, *The Geography of Childhood:*
        Why Children Need Wild Places

Deep below Lake Nasser in Egypt lie the homeland and sacred places of the Nubian people. A thousand miles north, in the middle of the Mediterranean, Turkish Cypriots and Greek Cypriots on Aphrodite's homeland fight over a land divided more than thirty years ago. In Cyprus, Egypt, and myriad places around the world, photos of lost family homes become shrines. Artifacts of daily living take on the patina of holiness as memories of familiar places, impossible to reach because they are underwater, across barbed wire, or torn down in the name of progress, become the seeds of nostalgia. For some, the displacement is an act of politics as one group in power moves another across the land. For others, countless immigrants around the world, the move is voluntary but perhaps no less wrenching, as the separation of person from place simultaneously severs and reinforces what it means to be home. To be home, to have a homeland, to be people who are *of* a place.

My ancestors were river people. I, too, was born at the confluence of the Hudson and East rivers near the northern tip of an island whose name in its original tongue meant "place of intoxication"—the isle of Manhattan. My mother's parents were born along the Neumans in Lithuania and my father's father along the Thames in London. I now live in a small house in Oxford overlooking the Thames, upriver from where my grandfather knew it, in a place where swans sleep at night looking like small, feathered icebergs on an inland sea.

The Thames moves with the slow rising and falling of sleeping breath. In spring, broken branches toss in rushing currents. Near summer, pollen paints long slow strokes of yellow and green beneath gracefully bending willows. In winter it crests in flood, lapping at the footpath. Always the river breathes with the sensuous motion of going out and coming in again. That tide, the pull in toward the shore and out to the distant sea, is a familiar rhythm. It lives not only in rivers and streams but in me as I move forward and back toward an ever changing sense of home.

A house is a dwelling, an *eikos*, a location, whereas a home is born in the marriage of imagination and place. In a world of constant motion and creation, homes are anchors of continuous possibility. Like wilderness and the wild, home is not simple or static but a system of dynamics moving in place, time, and memory, becoming a rich tapestry woven of location and movement, kinship and tradition, exploration and explication. As our primary storied landscape, home is the place we know so viscerally that we have difficulty separating self from other. For most, home is ambiguous, ambivalent, an ache. Like a feather drifting along in the current; the more we reach out for it, the farther it moves. We leave home, come home, and find that one vision of home exists in the cosseted safety of memory while another exists in an ever-present desire to be more so instead of less.

Home is too large a concept to explore in its entirety. Ambiguous, ambivalent, amorphous, the seat of self, of dream, of our understanding of place, home is the place of our stories, of who we are and where we want to be. A lost home comes with the sweetest candy-floss-flavored nostalgia; sometimes that same home is the one that also has the greatest power to make us sick.

My family members are not only river people; we have been immigrants for generations, defined more by motion than rootedness. The question of what makes a home, a homeland, is fueled for me by knowing that my roots are not so deep in any one country's soil. People of diasporas do not root in place but in each other, but that, like so many homeland stories, may only be a cultural myth designed to create a sense of identity. Homelands are all about identity.

The quintessential homeland conversation always begins with, "Where are you from?" Wherever I go, that question means something else. In the United States, where my accent locates me, I often have to ask if the question means where was I born, where was I raised, or what place have I just left since the answers are not all the same. In England, where my accent implies a larger geography than it does in the United States, the answer is broader. I am from the United States. If there's a request for clarification, it is that I come from a town "just outside New York." Sometimes I answer, in both time and place, "twenty minutes north of the city," which is helpful only to someone who can imagine the speed of traffic on the Palisades Parkway.

The ambiguity in the question "Where are you from?" stems from the English language itself because the very expression means that you are not "from" where you are now. To be where you are is out of your home. Modern day English does not allow the preposition with which the French take comfort: *of.* Lancelot du Lac was Lancelot of the Lake, and when asked in France where am I *of,* I do not answer in terms of location so much as ancestry and emotion. *Of* asks me "Where are my people and where am I home?" because when I leave that place I have only left it physically and am still possessed by it. *Of* is a statement of relation, *from* a point of departure.

Flying through Amsterdam or Brussels, my passport always invites the homeland conversation. While Kevin is usually being searched by armed guards for every last scrap of lint in his pockets, I am practically embraced. The expression is always the same. "Hello Dr. Van Gelder," they say with the strong Dutch pronunciation of my name. "*Welcome home.*" Welcome home? My branch of the Van Gelders left Holland for England 160 years ago. Three hundred years before that, they left the Cantabrian region of Spain, driven out by the Inquisition for the crime of being Jewish. In Holland, they changed their name and took on the same name as an entire eastern province, giving them anonymity and me the named identity of a returning member of the family. I am of Gelderland. Welcome home.

The embracing "welcome home" centers around the question of time. Is the United States my family's homeland because my parents and I were born there? Is England our homeland since my grandparents and great-grandparents were born there and I can find the houses where they lived? Holland? Spain? And before Spain somewhere else? And these are only the stories of one of my four grandparents' lineages. What of the rest? Do their homelands carry an equal weight internally or do I defer only to the line whose name I carry?

I am of immigrants; people who have moved from one place to the next, sometimes by choice, sometimes by necessity, but in each case carrying the familial memories and stories of another place. Like autumn seeds caught in the edge of shirt sleeves, falling out later in a new field and quickly taking root, we transients, who are defined by motion instead of rootedness, have a tangled relationship with the concept of home for we live in a space that tells us the place we are now is not our homeland but our homeland may never be home to us again. We are used to the mistrust afforded to weeds—even when those weeds become the deep pink fireweeds and wide-leaved chestnuts that will color the countryside in future generations.

In the Family Records Centre in London I pull old red leather books from numbered shelves to search for my name. No Van Gelders in the early 1840s, but by the 1850s I see Moses and Abraham Van Gelder's names recorded in smooth black ink. I know nothing of these people other that the places where they lived—their births recorded in red books, their deaths recorded in black books, their lives lived somewhere in between in a neighborhood known as Stepney Green. I am drawn to explore their neighborhood, to find the house where my great-grandfather was born, where he married a woman named Rose, where my grandfather Joe was born. In searching out the geography of my people, the feel of East London, the sense of the river, I look for something just beyond my grasp and yet I want to know something of who they were by knowing where they lived.

I am not alone in this feeling. The Family Records Centre is filled with men and women armed with organized color-coded folders, legal pads, some turned sideways with family trees hastily scrawled, others working on complicated charts that look like Churchill's World War II naval battle plans. In every corner is an angry old woman who has waited a half hour for the November 1887 record book and reaches out to grab it as you replace it on the shelf. She makes an unmistakable sound of someone who believes her ancestors have been violated by having to share a volume of birth records with yours. These are the people digging into the past to find something, find someone, find who they are by tracing the lineage back as far as they can go. They, and I with them, are searching for pieces of a very personal story. Who we are is who we are *of*. Finding the threads of that story and tracing it through the annals of time are part of the personal odyssey of finding home.

Simon Coleman and John Elsner in *Pilgrimage Past and Present* suggest that odyssey stories speak to the transient's desire for the embrace of a homeland. "Like pilgrimage, the voyage of Odyssey became in the ancient allegorical tradition (both pagan and Christian) a potent symbol for the inner spiritual journey to one's authentic home."[1] While Homer's *Odyssey* can be read as a multilayered story on the nature of hospitality, hubris, and heroism, Coleman and Elsner see the *Odyssey* as the archetypal story of the search for our "authentic homeland." Like Goldilocks trying out all the wrong beds, Odysseus tries out all the wrong beds, too, always trying to get to the one that he calls home. His son Telemachus leaves home to search for a father he has never known, except through story, with the hope that his father can restore their family and household to its authentic state. One is the journey from the outward, of Odysseus a changed man trying to get back home. The other, Telemachus's, is a journey from a home to learn what self is. Both stories have at root a desire to harmonize the concepts from and of. Both speak of being indigenous to a place, to being rooted.

Indigenous knowledge is fed by the storied cycles focusing on the processes of adaptation and harmonization. Adaptation stories tell the tales of a people learning to live in an environment. These stories include floods and hurricanes, attempts to grow sunflowers and discovering too little sunlight, or epic battles between suburban homeowners and a family of masked raccoons who systematically raid the neighborhood's garbage cans with great whackings and clatterings.

In time, newness is replaced by a code of adaptive behavior, a "way of being here." The harmonization stories that arise in this period become explanations for ways of dwelling. They say that people do not put out their garbage cans in this neighborhood because of the raccoons, or why no one has a garden in the patch of soil that floods every spring. These stories, the ones that move from tales of trial and error to the development of a social code become embedded in the way in which we live with a place. In that, we become part of a place by knowing its nature, and in return its environment shapes who we are. Keith Basso suggests that the place co-creates with people: people shape places and are shaped by them; raccoons rummage in garbage cans at night so now cans are never left at the curb. These are small ways of adaptation. If our families have been in a place for a long time, we feel a tighter bond and a surer sense of self in harmony with the environment in part because we have inherited the stories of adaptation without having had to experience the conflicts ourselves. "What people make of their places is closely connected to what they make of themselves as members of society and inhabitants of the earth, and while the two activities may be separable in principle, they are deeply joined in practice," writes Basso. "We *are,* in a sense, the place-worlds we imagine."[2]

Place becomes part of us because we have made adaptations so we can live harmoniously with it. In time, who we are is shaped by where we are *of.* These ways of living harmoniously with a place are largely unconscious. It is not that we note their presence so much as we note the absence when we move to a new environment. At home, we find familiar comfort in the colors of our landscape, the slant of afternoon light, the sound and shape of wind. We know intimately the "still-eyes-shut" sounds of morning. Blackbirds singing morning greetings. Garbage trucks wheezing along the curb. The heavy drumming of summer rain.

When I first moved to England I spent months feeling unsettled. The light was all wrong, the clouds moved by too quickly, and rain was not a question of if but when—usually precisely when I had left the house without an umbrella. I noted again and again, usually with very wet hair and sodden shoulders, that I lacked the ability to predict local patterns. I was expecting the landscape to conform to me.

In time, I bought a small umbrella to keep in my pocket and started dressing in layers. Soon I discovered that no rain ever lasted more than a few hours. It was bound to pass, unlike the rain in New Jersey, which usually stuck around all day. As I stayed through the seasons, I started to know when the chestnut trees would be in bloom, where to find the best blackberries, and that I could count on the first snowdrops to emerge in Christ Church Meadow at the sunny patch near the curve in the stream by the first week in February. Comfort, a feeling of belonging, came through adaptation. Adaptation led to expectation of patterns. Now I look forward to February, wondering when the first snowdrops will unfurl to tell me it is spring. Will they come up where they did last year or someplace new?

Like trees with radiating rings moving outward, recording the passage of seasons, we come to feel at home because the patterns of life, adaptation and prediction, become ingrained in us over time. We develop a storied sense of place. Paul Gruchow wisely tells us that "a home, like a garden, exists as much in time as in space. A home is the place in the present where one's past and one's future come together, the crossroads between history and heaven."[3]

Home is our most primary storied landscape, the one

with which we have the deepest roots and to whose story we continually add. But those stories are not necessarily the dramatic tales of adventure we acquire on our travels. Home stories are more like varnish brushed onto a newly sanded floor. Invisible at first, but, layer upon lived layer, they eventually make a surface we can see of its own. One that is both protective and hard.

The philosopher Gaston Bachelard, writing in the early 1960s, was one of the first authors to explore the creative-emotive power of home. In *The Poetics of Space,* Bachelard posits that home is "the seat of our reveries" and to possess a life of creative imagination we must have a memory of the imaginative power of our first home. "The house we were born in is more than an embodiment of home," he writes, "it is also an embodiment of dreams."[4]

Edith Cobb, writing in *The Ecology of Imagination in Childhood* in the late 1970s, offers a further understanding of the spiritual dimensions of childhood and place. She believes that the adult need not revisit the childhood place as much as that the person must have the lifelong habit of bringing together the power of creative wonder that is forged in the childhood experience of place and time. "The sense of wonder is spontaneous, a prerogative of childhood," she writes. "When it is maintained as an attitude, or point of view, in later life, wonder permits a response of the nervous system to the universe that incites the mind to organize novelty of pattern and form out of incoming information.[5]

Bachelard believes we need to actively revisit our relationship to our first home if we are to maintain a sense of *poesis* or a creative relationship with the world. According to Bachelard and Cobb, we need not revisit the first home itself but can revisit the emotions of the special places of home— the smallness of nests, the mystery of attics, the window seats of daydreams—to find our creative spirits alive.

Replacing Memory

Bachelard, Cobb, and Gruchow all suggest that home is a place known in time, space, and story. The nature writer Barry Lopez refers to the story developed from repeated encounters with the same places as "replacing memory." Lopez believes that if we return to the same places at different points in our lives we form a storied connection to those places and develop a continuous narrative structure. That narrative structure creates an opportunity to reflect on the ways in which we have come to know ourselves. Places and repeated encounters with them offer us an opportunity to develop animistic, live relationships with our environment, a narrative framework for our own selves, and direct access to the process of time.

When I was four years old, I visited Tsavo National Park in Kenya, a place where the elephants that roll in its unique red clay soil come to look like small red mountain ranges migrating beneath the shadow of Mount Kilimanjaro. The lodge has a lovely stone patio where people gather in the evenings to watch the animals coming in to the salt lick and water pool. I have seen two-day-old elephants there and tiny rock hyraxes, their closest genetic kin.

As a four year old, I struggled to see over the stone wall surrounding the patio. One of my strongest memories from childhood is the feel of the stones in my hands as I tried to boost myself up followed by the magic moment when I could finally see animals illuminated in the pool of ambient light below. I was terrified that in my leaping, I would jump too fast or too high and would land directly in the salt lick. It was a very powerful fear. It left a visceral memory.

Ten years later, I returned to Tsavo. While most of my memories of Africa had been colored by seeing slides from our trips, my memory of the high wall of Tsavo lingered in my hands. We pulled up to the lodge, and I immediately walked through the dining area to the porch.

"Have they renovated?" I asked, seeing no sign of a high stone wall.

"What? No. Same as always." My father gestured to a favorite sign indicating that people feeding anything to crocodiles would be required to retrieve those items.

"But where is that high wall over the salt lick?"

He laughed and led me over to the edge of the porch. It came up to my waist. It had not changed at all. Only I had.

While I only visited Tsavo three times in my life, I replaced memories each time. If the motion of replacing memory makes us revisit the first time we encountered an experience or location with each return, then we exist in a harplike state with our first homes, making quiet resonating strings back and forth across the space of place and time until we may look internally like the graceful strands of the Brooklyn Bridge or a nest of impenetrable knots. Whether consciously or unconsciously, like the tides, we are always leaving, and always going home, leaving invisible strands of connection behind.

Homelands

Homelands are forged in the moments of conflict when a shared set of stories brings together a disparate group of people in the face of an enemy to feel as though they are one. Usually they are united by a set of stories that speak to having an intimate knowledge of living in a place or an intimate knowledge of the history of that group. That past is only accessible to its membership, and it is through that inaccessibility of holding the past, and having seen a place change into the present, that the evocative power of home-lands rests.

How we come to feel like inhabitants of a homeland is through knowing a place through a period of change. That transformation carries a storied understanding of a time "before," one to which we can never return, and a very present "now." Our conscious participation in irrevocable change makes us intimate, indigenous members of the in-group in that place.

Recently, I went for a bike ride through my old neigh-borhood with my friend Evelyn. She and I have known each other since I was four and she was five. We are the keepers of a lifelong collection of each other's stories. My first memory of her is a summer's day by the town pool and Evelyn pa-tiently teaching me how to dive. She was older, wiser, and more experienced in the ways of the world. After all, she was five.

On our recent neighborhood revisit we rode bikes, just as we had as children, and traveled a familiar geography that we had known intimately in the years before we could drive. We immediately found ourselves in the liminal space of the storied landscape. Simultaneously both ten-year-old girls and women in our thirties, we found that our replaced memories flowed with a continuous narrative patterned in the ritual of evoking people, place, and story. "This is where . . . lived. Remember him? He . . ." House upon house, we rode to the sacred spots where we fell off our bikes, once caught tadpoles at the pond (now long dried up), fished in the stream by the old swim club (now too toxic for any ani-mal life because of lawn fertilizers), and played. We avoided busy Lafayette Road out of instinct, because Evelyn had never been allowed to ride her bike on it, and the "new" Pondside Park, built almost twenty years ago, because we had no stories there. Instead, our bike ride focused on gauging the passage of change during the period when we had been witnesses to it.

We could not go back to those halcyon days of the early 1970s, nor in reality would we want to, but in sharing the stories of the place evoked by the feeling of wind in our hair and the feel of bicycles on well-worn roads, our little town became both the town it is now and the town housed in a forever unchanging place—what Bachelard calls "the land of Motionless Childhood."[6] The same place that as teenagers we couldn't wait to leave is the one to which we strangely can never find our way back. It is the place of our first dreams, our primary myths about who we were and who we would become. It is against those dreams of who we were to become that we still measure ourselves. In those foundational myths we find our own truths, and as old friends we hold those dreams for each other when one or the other loses her way.

In sharing our stories we were also able establish a ring of identity that says we were residents of Harrington Park, New Jersey, in the 1970s and 1980s. Those who came after us, ones we think of as newly arrived, can't know it in its old incarnation. As with all peoples, our history begins with our arrival. But, having been present for a period of measurable transition, we can behave like a grandmother at a favorite grandchild's wedding, saying to any and all who will listen, "I remember when she was only a wee thing."

I remember when the Colorado blue spruce in the yard was only my height or when there were lilac bushes along the edge of the property before the maples got too tall and blocked out the sun. "I remember . . . I remember . . ." sounds the clanging bell of a homeland. At a community gathering, the "I remembers" ring like a carillon of bells, story upon story being retold. These are stories of identity and kinship and stories that make the past inaccessible to the new in the present. Simultaneously celebratory and insulating, the "in-group" celebrates a knowledge inaccessible to the outsider. Storied time holds the keys to access, but the process of time allows that newcomer to become part of the old guard, too, in the face of the next wave of people to come through.

The desire for an "authentic homeland" is about wanting to look around and find ourselves surrounded by kin—the people of whom we are a part—with faces like ours, eyes, hands, the same names and language. More important, we want to be with people who share the memories of the same set of stories, our stories. Our continuum ensures that our people will exist after we are gone and that we are carrying their nature in us. We want the senses of being *of* and *from* to be one.

What if the opportunity to replace memory is denied and there is no way back to the touchstones of memory? For displaced people the inability to reconnect provokes a range of passionate feelings, from the homesick holiness of nostalgia to more extreme forms of violence and hatred. Homelands and the denial of access to them are at the root of many of the current conflicts in the world. Is it because we, as humans, cannot be severed from the places that matter to us? Denied access to the places of our stories, do we become our most extreme? Whether immigrants, migrants, refugees, transients, or homeless, the causes of displacement differ for each individual, but the questions surrounding the displaced person's relationship with place become much the same. What happens to people when they are cut off from the possibility of returning to a place they call home? What stories grow in that space? What impact does the feeling of placelessness have on the imagination? Who are we when we feel that we are denied a homeland? Are no new roots possible?

In the spring of 2000, I had the opportunity to spend time in Cyprus where the questions of homelands and displacement are not abstract concepts but part of daily living for the thousands of people who live in Nicosia, one of the last divided cities in the world. I will share the stories I learned there as they speak to a relationship with place few in the United States know as intimately, the roadblock to going home.

Aphrodite's Divided House

On my first morning in Cyprus, I stood in a bank parking lot in the center of busy Nicosia. Still jet-lagged and bleary-eyed from the late night flight I had to look twice when I saw Turkey's flag waving across the road from the bank, right in the middle of the city. Searching the sky, the blue and white Cypriot flag waved from the other side of the same road. Crescents and crosses hanging in the sky, both symbols woven on cloth, the flags waved in the dead space over the "Green Line" symbolizing the painful divide of one country from another drawn right through the capital's heart.

Across the street, the faded blue and white sign designating the protectorate of the United Nations was half-hidden in a grove of overgrown palms. This was not how I had imagined the Green Line, but when I saw its sign marker I was transported back to Kenya in the summer of 1973, when I dug in the ground at my feet hoping to unearth the black line of the equator, which I was sure must run just beneath the surface of the red clay soil. On the globe it was so clear. Why was there no line running along the equator so I might know the northern hemisphere from the south? The map of Cyprus, purchased at 3:00 am in the airport duty-free shop the night before, had the words "Territory Occupied by Turkey" written in bold black letters across outlines of mountains, beaches, and olive trees. This time I knew better than to dig in the dirt to find those letters. Through barbed wire, soldiers, and blue signs marking a space known as the dead zone, I could see what those letters really spelled.

I had gone to Cyprus at the invitation of two friends. One a doctor, the other an educator, both so dedicated to the process of peace that they have often risked life, reputation, and their family's safety to find ways to bring a peaceful resolution to the "Cyprus Problem."

When we met the year before in graduate school, I had to confess that I didn't know for sure what the Cyprus Problem was. I thought it had something vaguely to do with Russian missiles or poor date production. They spoke passionately of place, of story, and of the transformative power of home. A year after we met, they invited me to come to Aphrodite's brokenhearted island to see what it meant to live in a house divided. My friends wanted to know how to build the road to peace for people who had made the story of their displacement holy.

On my spring day in the bank parking lot with both flags in view, I could not help but think of the word *history*. During Women's History Month in March it had been common to see T-shirts with "Herstory" written on them. In Cyprus, standing in the empty space between two flags, the question of history was the key to the perpetuation of the stories of people and place. Whose stories were perpetuated? Which story is told? Geographically, Greek Cypriots and Turkish Cypriots may claim the same longitude, the same latitude, the same Mediterranean sun, and the same legacy of Aphrodite, who was said to have washed up on the western shore, yet across that barbed wire they were worlds apart.

"Here, I cannot go anymore," my Greek Cypriot guide, Eleni, told me, gesturing to a sign that read "United Nations." Next to it, were posters of a couple who were presumed murdered four years ago by Turkish Cypriots. Their wedding picture shone back at me, filled with expectation and celebration. This was one side of a story. I am told by those who have crossed over to the Turkish side that pictures of murdered Turks decorate the other side of the dead zone. The pictures, like the three-headed Cerberus of Greek myth, serve as guardians to the world of the dead, perpetuators of the memory of bloodshed and keepers of the hate.

Next to the poster board was a black cutout of the island of Cyprus with red paint splashed like blood across its border, barbed wire twined savagely across the area now occupied by Turkey. Near the entranceway to the dead zone was a compound that was once the Ledra Hotel. A group of uniformed schoolchildren had gathered in blue uniforms in the cool archways, each holding a red flower. They listened to a speaker whose voice was broadcast over loudspeakers. Although I could not speak Greek, Eleni told me that his staccato bursts of strong language were words of war. "Never forget your homeland." The words drifted over the children and out into the dead zone. "The children," Eleni told me, "are here to learn their history." What do these children learn in school if this is where they go on a class trip?

In countries where there are wars, there are always war heroes. Every village in France has a central column or a weeping angel carved with the names of young men who are the stony generations of the dead. In the villages of Cyprus, I see monuments shaped like young men in soldier's garb. Tall cypress trees on either side of these permanently petrified youths underscore their rooted connection to the island, its name. I have to ask which war left them with their youth entombed in stone.

"Do you think they represent the hidden war in which Greek Cypriot killed Greek Cypriot? Neighbor against neighbor?" my friends reply. That is a story no one tells. No one has rushed to build statues of the grieving mothers still dressed in black. To keep the anger pure, there can only be one set of heroes, one war, one injustice. Cyprus has been a nation occupied by larger empires for the last five hundred years. First came the traders of Venice, then the Ottomans, then the English, and finally, left to create its own independence in the 1950s, it tore itself apart. But those stories are not the stories retold on the monuments and in the speeches. The enemy that time lived too close to home.

One friend's defining moment came when, as a young boy, he witnessed a political assassination. In the 1950s, after the British left Cyprus, hungry people filled with suspicion and anger turned on each other in a civil struggle for power. His father owned the village cinema. One dusty afternoon as he sat in the back row watching the feature, a man walked in and sat two rows in front of him. Two others came in, sat behind him, lifted a gun and pulled the trigger. The assassins had once been friendly neighbors of the dead man, who died for the sin of handing out the opposition's political pamphlets. These spoils of war, the blood on the shoes of a nine-year-old boy, became his lifelong trail to peace. Never again, he swore, never again.

He laughs when tells me that he once asked these questions at a televised conference filled with powerful nationalists. "Why don't we have peace heroes? Why don't we celebrate the peacemakers with monuments and pictures? Why is their work not the stuff of history?" They could not answer him. Instead, they stopped inviting him to conferences, especially ones on television.

Eleni takes me from the Green Line and Ledra Hotel

into the heart of Nicosia, Ledra Street. Ledra Street is the Cyprus of imagination—an enclosed walking mall lined with shops and twisted alleyways where singers armed with cheap Casio keyboards play Greek renditions of old American jazz songs, painfully. Yet in the heady fragrance of the Mediterranean air and tables littered with crimson roses it is somehow romantic. Eleni shows me the ancient outer wall built by the Venetians, who occupied Cyprus before the Ottomans, before the British. She tells me some of the oldest ruins of Mediterranean civilization still shine in the Cypriot sun. Storied landscapes.

On Ledra Street Eleni passes friends and old women, people out taking in the day, window shopping. I am lulled into a false sense of security, feeling as though I am in Cambridge or Bordeaux as the shops are filled with familiar names. But the street does not end in a wide boulevard or a taxi rank. Instead, we hit the wall. There it is again. Ever present. Divided.

Peeking through the fence, we can only see the wall on the other side. We can go no farther. No shops leap across the dead zone. An open storefront is plastered in photos—faces, houses, beaches Greek Cypriots can never visit again. Captured on film they become burned into memory.

"Of course I hate the Turks," Eleni says matter-of-factly. She and I are the same age. She was five years old when the invasion took place. In the summer of 1974, Evelyn was teaching me how to dive at the pool. Eleni learned other things. "My grandfather cried every night of his life for never being allowed to go back to his land. Do you know what that means to separate a man from his land? I didn't understand what that meant when I was younger, but now I am starting to. Come, let's go look for a pair of sunglasses for you—"

Her hatred is offhand and assumed, evincing the power

found in the assumption of a mutual shared hatred. I thought of the children at the Ledra Hotel learning the rhetoric of hate. Will they associate that language with the lovely red flowers they held in their hands so that they become one and the same?

Once in Egypt I had a similar moment. Outside Abu Simbel, Ramses's smug monument to the conquest of Nubia, Kevin noticed that our guide, Ahmed, was wearing a Boston Red Sox cap.

"Do you like the Red Sox?" Kevin asked, making polite conversation.

"No. If I had known what this was, I never would have bought it. I hate America and all things American." He wore Levi's jeans to accompany his Red Sox cap.

I got very quiet.

"Why?"

His litany was not unfamiliar, and many of his political complaints weren't too far from mine. Ahmed's words didn't scare me as much as the ease with which he spoke. We were in a British tour group, and he assumed he was among friends—certainly we all hated all things American. No need to hide it. In Abu Simbel, with the Nubian homeland drowned beneath our boat, I asked the same question I had been asked in Cyprus. Who builds monuments to the heroes of peace? Do they just carry the blood on the bottoms of their shoes? Why has Ramses II's monument to oppression lasted four thousand years and remained one of the great tourist attractions of the world?

In the evening, after my tour of Nicosia, my two friends invited me to a coffeehouse in the old section to see the bicommunal youth group they organize. We navigated winding streets with colored doors and ornate wooden gates. The bicommunal youth group met regularly in this coffeehouse in the oldest section of town. Minutes after we arrived, a young teacher stopped in briefly with a roll of posters beneath her arm. She unfurled one for me. Across the wings of a white dove written in Greek and English were the words "History. How should we teach it? How should it be taught?" I remembered asking my own students this question just a month before standing in front of the Liberty Bell in Philadelphia. Who decides what is our collective history? We never came up with one answer in front of the Liberty Bell. I would not expect these young people to do so either.

The young teacher left after wishing the students well. In their closed room in the coffeehouse, the youth group tallied the responses from a survey they had conducted regarding student attitudes toward bicommunal rapprochement. These surveys had been conducted by young people on both sides of the Green Line. The responses, so candid, admitted to the problems and seemed so optimistic until the last question.

"When do you think these issues will be resolved?"

Again and again they tallied marks from the box that read "Never."

Victor, the seventeen-year-old leader of the group, explained to me in crisp English that the group was in the midst of planning a bicommunal conference and needed to decide who should be invited as the keynote speaker. President Clinton? They looked to me as if Bill and I played poker together on Friday nights. They then suggested the president of Greece as a strong possibility. Intuitively, they comprehended the gravity of their work. While they discussed their classmates' responses to the rapprochement surveys, they commented that none of their own surveys ended with the word *never*.

Conversation jumped back and forth from Greek to English as they invited me to participate. In between heavy conversations about peace work and hopes for change, they teased each other about soccer championships and their love lives. These young people were no different than young people around the world, and yet they were. They had the power to change "Never" to "Someday." The beautiful young men I saw laughing over their rival soccer teams did not need or want to become monuments to war, but they had to fight to be agents of peace.

As the meeting came to a close, my friend addressed the group.

"You did not ask to inherit this problem. You were not born when the invasion happened, and yet you are being asked by the preceding generation to solve it. You can decide what to do about it. What will you do?"

They asked me to pose them a question. Two months before, in a homeless shelter in Hoboken, I had asked the question, "If places could speak, what stories would they tell?" Knowing that the homeless often interact with more places than people, their stories spoke to wrenching memories of place and loss. Stories held their identities like anchors at times when they could not hold on to their selves.

In Cyprus, where the history of two groups of people was so deeply tied to a geographical location, I asked that question again. If these places could speak, what history would they tell? Could the limestone cliffs tell the difference between Turk or Greek? Did Aphrodite really wash up on those shores? Who would claim her now? At the Green Line, where people could not pass to visit the places of their own family history, I wanted to know if the stories die when no one can go home. Who decides the border of a homeland?

When people replace memory they have the opportunity to revisit themselves because they experience the compression of time through the telescoping power of place. When they cannot revisit the place, the very action of being permanently severed from it can create a profound sense of nostalgia that exalts the place from the mundane to the holy and the source of a perpetual form of unassuaged grief. I know this from the history of my religious background: people exiled become the singers of grief until at times the grief seems larger than life.

Almost the entire homeland of the Nubian people lies in a watery grave beneath Lake Nasser in Egypt. At Abu Simbel our guide Ahmed told us how much happier the Nubians are now that they've been relocated to Aswn.

"New houses. Better jobs. Big money. They've made out much better than regular Egyptians."

Underwater. Graves of grandmothers. Holy places. Temples moved to higher ground—saved but only paying tourists to visit them.

"Some Nubians make a lot of money recording music. All sad songs. Don't believe them. They make more money that way."

Our Nubian friend, Ibrahim, took us to his village above the first cataract in Aswan. Houses are blue and yellow, and children chase after us stealing pens from Kevin's pockets. Looking at the women in black lace dresses, Kevin asks, "How do the people feel about the lake?"

Ibrahim is slow to respond. It is not politically correct to speak his opinion. His black hands run slowly down the length of his white galabia. "Kevin, how would you feel if you could never go back? You know what is in my heart. I do not need to tell you that all of our stories have no home. You feel this in your heart."

One of my Cypriot friends was sixteen years old and on a camping trip with his girlfriend in the southern part of Cyprus during the invasion. For three months he lived in a refugee camp, wondering if he would find his parents again. It took years for him to accept that he could never go home again. How could some other young boy have his room? Did they still use his family's furniture—a three-bedroom house fully furnished in move-in invasion condition? Who moved into that Turkish family's house in the south? If those places could tell stories, would the new owners recognize the characters in them?

The word *nostalgia* comes from Greek. Perhaps it is fitting in Cyprus's climate of grief and love to speak of nostalgia, and yet nostalgia is a dangerous brew. When people are blocked from returning to the places where their stories live, they develop a nostalgia that elevates those places to the holy. Old homes, bicycles, and pictures become more than the stuff of everyday life. They become the icons of a religious desire to reconnect to emotional landscapes. Nostalgia feeds war as it is the ambrosia of the emotional landscape. One broken bicycle from the past, one picture, one glimpse of home—and some are willing to die to have possession of those things, those places, again. Paul Gruchow writes:

> Nostalgia, we believe, is a cheap emotion. But we forget what it means. In its Greek roots it means, literally, the return to home. . . . Nostalgia is the clinical term for homesickness, for the desire to be rooted in a place—to know clearly that is, what time it is. The desire need not imply the impulse to turn back the clock, which of course we cannot do. It recognizes, rather, the truth—if home is a place in time—that we cannot know where we are now unless we can remember where we have come from.[7]

Gruchow's nostalgia seems to be that of the adult yearning for the memories of childhood and home. This nostalgia is different from the one in which the division between person and place is not so extreme. If we are reflected in the places we call home, are we lost when we are displaced from them? To choose to leave is to make peace with the process of change. That is why immigrants have a word whose root is the word *migrant,* which comes originally from the Latin word for mutation or change. But to be displaced, like water suddenly splashed from a glass, or suddenly driven from a place does not permit the process of transformation, grief, or separation. Instead, there are perhaps only holes, and those holes are easily filled with grief, anger, and a holy nostalgia for which there are few cures.

Outside the Nicosia coffeehouse in the darkness, with the stories of what the land would tell if it could speak, my friends and I spoke about history. They told me of their experiences in working with Turkish Cypriots. The historical events of one community do not match the historical events of the other, although they occurred in the same places.

"It is like they are from two different planets." My friend pulls his hands apart, making a wide gap between them. His long, tan fingers point toward each other.

"And yet, when the events are put together, this happens." He moves his two hands and nestles the fingers so they form a whole.

Each hand is an arrow of events of its own. Together, they are one. They are full. They are Cyprus. The connection points link. It is a powerful image. It speaks to me of place, something you can hold in your hands and cannot slip through. Whole.

As I head to the airport, my friend tells me one last thing, which will stick in my head as I travel across the time zones back to my own country.

"For as long as people believe only one view of history, they will travel in a continuous circle, wearing a deep rut, by its own action, building up walls. It is only when the different groups of people who share a history come together and see the ways in which their histories intersect and interact and connect that there will begin to be something else. Once they can resolve their past, then, and only then, will they be able to create a new sense of history and move forward again. Their feeling of being displaced must be honored. Their stories must be heard, but they must build new stories, too."

Nationalism feeds on nostalgia, allowing the politics of place to become a religion bred on the perpetuation of stories that cannot be replaced. Because they cannot be replaced, they live only in people, taking on a mythic status and a power so strong that the retention of those stories, those borders, become barriers because they are rooted in an inaccessible, immobile past; one trapped under glass in a museum instead of being able to migrate, mutate, and change organically into something new. If displacement finds its strength in the past, there must be a way of making the past a conduit to the future without the link to place. But is that possible or are we, as humans, too place oriented to be able to sever those emotional ties with ease? Are we as rooted as the trees, and as dead when we cut our roots, or are we seeds with the memories of the old trees stored within us, ones that can replant, find new soil, and begin again?

In the spring of 2003, for the first time in almost thirty years, the passageway known as the dead zone opened up along the Green Line. Hundreds of people crossed through the barbed wire gates, clutching photos of their homes, their families, themselves as children, able once more to touch the places that had once been their homes. It has not resolved the political conflict between north and south, but perhaps for the people who could not replace their memories it will allow them to make peace with the past as they will face the glass box of their memories and see that, although those places are frozen in memory, they have thawed through time and become someplace new.

FIVE

At Home in the Alchemy of Love and Fear

All inhabited space bears the essence of the notion of home.

GASTON BACHELARD, *The Poetics of Space*

On September 12, 1986, my mother wedged her midnight blue Plymouth Reliant K-car into the only empty space in the garage, gunned the engine, and filled the tight space with carbon monoxide. My father found her some hours later and phoned the police who pulled the car, and my mother's body, from the garage, leaving us with only empty space.

In the year after my mother's death, my brother drove that same car back and forth to college until it, too, died, this time on the I-95 approach road to Philadelphia. I never asked what happened to the car. In my imagination, the car broke down and my brother walked away, as fast as he could, no longer having to sit in the same seat where she had taken her last breath. Free.

I know this is not how the story goes, but my knowledge of home has been about learning how to create a home in a space that has been filled with life, decay, and death. It is fine to abstract these concepts when talking about a forest floor or an ecosystem but so much harder when it is a room in your home and the place of childhood dreams.

Although the term *mother* is so often attached to the embrace of nature, my mother was my wild. When I was a small child, she was my refuge, and the transformation from safety to wildness is a motion I came to know intimately. Like all of us, perhaps, my mother was a jewel of many facets. I knew her as one face; the people in the community knew her as another; my father knew her as another still. She spun and spun, often slowly enough to catch the light from which she glowed. Sometime in the early 1980s, she lost the capacity to slow the spinning and instead, fueled by a growing mental illness, spun with abandon simultaneously inward and farther and farther out of control.

In her spinning my home lost its emotional center and became instead a cacophonous fugue. There was no safety in instability, no pattern to predict; life was lived moment to moment, mood to mood. In the four years of her illness wild places away from home began to become my safe havens and home the most dangerous environment I knew. I never spoke of it. My home and my mother were my secret shame. For it and her, I had no words. My wild.

On the morning after she died, my father and I cried together at the kitchen table. We both knew so much about leaving; we knew nothing about being at home. We loved each other and had a friendship with roots as deep as ancient oaks and in that we knew the feeling of home, but we had no knowledge of being rooted to a place we loved. We believed in motion. We had never truly unpacked our bags.

For the next eight years my father made attempts at learning how to be at home, but a lifetime of leaving made the staying all the harder. In the winter of 1994, amid the worst ice storms of the century, he developed acute leukemia and died just two months after turning sixty-five. Later I found the Valentine's card he never had the chance to send me buried beneath the mountains of books, newspapers, and cereal bowls on the kitchen table.

At Home in the Alchemy of Love and Fear

In his death, I knew homelessness. I lost the only person who knew me plain and loved me anyway. Although his house was an uncomfortable space filled with old knitting projects and fish tanks rotting with guppy experiments gone awry, his laughing eyes, warm hands, and broad mind had always been the place I had found home. He and I were made of the same material, and that resonance made our relationship easy. At his funeral, the rabbi pinned black ribbons on us, which were then cut to symbolize the severing of the relationship. I did not need the symbolic gesture. I already knew what it felt like to have the material from which I was made suddenly ripped in half.

Home is born in the alchemy of love and fear. From the very beginning family shapes the architecture of existence. It is those people who see themselves mirrored in us—the parents and grandparents, brothers and sisters who are looking to see whose eyes and hands we have—who recognize something so fragile in our sweet milk baby smell that it compels them to lean over and protect us from all that they fear. And, like masons crafting ancient domes, they shape the space of love as one hard and protective on the outside with a smooth arc of safety on the inside. This motion of shaping the environment bred of a fear of outside forces—the putting up of bars on the crib so the baby doesn't fall out at night, whispering so as not to wake a sleeping child, covering, and cosseting—are all part of the tangled architecture of home where the lines between self and other blur. When the combination in the alchemy feels right, those walls form a lifelong sense of emotional security from which to explore. When the alchemy becomes unbalanced, those walls can create a prison, an impenetrable fortress built from a maze of other people's fears.

We can no more know the right alchemical formula for turning lead to gold than we can know the perfect balance of love and fear required to forge the ideal sense of home. Too many elements, too many factors. The sensitivity of that alchemical combination serves as the base formula of a life-long set of questions. Who am I? Who were my parents? Why did they paint my childhood bedroom pink? What did they believe? What do I? In the home environment, like no other place, person and place merge. We are both *of* a people and *from* them. Impossible to separate, we dwell in each other.

In-Group—Kin Group

One of the most powerful phenomena in human social development is the concept of the in-group/out-group. As we come to dwell in each other we form protective relationships with those with whom we have greater and greater degrees of connection. Much like when a leaf falls in a river and waves radiate out from it, forming wider and wider rings until those rings reach the banks and ripples then move back inward again, we interact with those around us, forming wider and wider rings of identity and belonging until we hit walls. At the very center, we find only ourselves. Close in, our family, our kin. Farther out, the rings include those with whom we share a sense of geography, a collection of beliefs, a similar worldview, the same taste in music, stories. At the outer edges the rings appear like levels of Linnaean taxonomy. Apes are in my primate family circle while hamsters are not. But hamsters are mammals and in my in order while snakes are not. This works until I try to decide if I am closer to planaria or python. When someone might arrive from "outer space," the planaria and I might find we have much more in common than we originally thought.

All in-groups are based on commonalities in the face of differences. We humans, as pattern meaning makers, continually look to taxonomic systems as means for guidance in deciding safety in relationships. With whom we can identify in a given moment is a key determinant in survival.

Unlike the leaf in the river, which creates the impetus for the initial ring but has little control over its farther rings, our rings of allegiance radiate and contract based on what we confront. Fear, the great sculptor of the exterior boundaries of home, makes us constrict the bands and move tightly in to the closest ring in which we find safety. Unthreatened, we share the pleasure of developing wider and wider rings. Threatened, we can become, in an adrenaline-fueled instant, all of the qualities of the safest ring of protection as we look immediately to others at that level for support.

Home is forged in that tangled alchemy of love and fear wrought by those who believe we are of the same material as they. From the very beginning, those in the closest band of our in-group, our family, shape the space around us. And that space contains a rich tapestry of the sacred, the mundane, and the profane all woven into one.

The Sensuous Landscape of Home

We learn of home through hand and mouth long before we have words to speak of it. As a baby, it is a groping and exploring that teaches us that there are soft things around us, things to touch, to taste, to feel. Of all our sensations, the most powerful and primary is touch. In English the expression of emotion is the same word for the expression of touch: *feeling*. When moved emotionally we speak of being "touched." At root, we are sensual beings who learn first boundaries of where we end and our mothers begin by touch. Hand and mouth, seek and find, drink in, cry out, nourish, sleep. Home becomes the softness of warm body enveloping and feeding. The newborn already knows the familiar heartbeat it once heard like regularly chiming cathedral bells from the inside, now muffled on the outside. The same voices fill the air with vibration. This is the familiar and the familiar in the body of home. Family.

Those little hands learn to touch things, feel the texture of a blanket, of a breast, its own soft skin. That feeling of the world around becomes the first level of acclimation. To become familiar to ourselves, to know how we feel, to know how our mothers feel; we cannot separate our sense of home from our sense of family because at the very beginning they were one. Our place and story were one.

Person as Place

While we are accustomed to thinking of places as locations, we would be wise to think of people as places since our first place, our primary home, was our mother.

Imagine sending mail to each other: "To Leslie Van Gelder c/o Rosalind Van Gelder's Womb." Where she moved, I moved. Inside her, in my own little amniotic sac, I was dependent on her and yet I could move separately enough to offer up an occasional well-placed kick to the bladder. I could move inside but not outside of her, and through my mother I experienced the great umbilical conundrum: the borderline between self and other.

The beauty of thinking of ourselves as evolving places is that even from the very beginning we have been in and experienced an environment that no one but us has shared, and yet that environment is someone else who also experienced us. Perhaps, like the double-woven basket, we have developed a relationship that by its very nature is both inner and outer. In the very first moments of life we have had to know that the source of life is an external connection to others, but our understanding of our experience gives us a distinct point of view that only we know through memory. We have been *of* and *in* at the same time. Like places we visit over and over again, we ourselves are storied landscapes, where each scar holds a story and each moment hangs pregnant with possibility. As humans, we are always in the process of storying because we take in our world and we try to make meaning of it. The combination of experience and reflection allows for story. Even, perhaps, at birth.

If doctors could conduct exit interviews with newborn babies, they would hear stories of climate descriptions and complicated favorite dance moves and kicks. Perhaps descriptions of a great love for garlic or an equal hatred of Old El Paso salsa. Doctors might hear of frustrations over the sounds of a dog barking or the love of the rhythm of a mother's walking when the baby wanted to fall asleep.

From the very beginning of life, our experiences with our very first environment helped to make us the unique individuals we are. We are not factory-produced tabula rasae; we are already, at birth, experienced beings with stories to tell, just, perhaps, not the ability to articulate them. For the remainder of our lives we will be both shaped by the environment from which we derive life and perhaps a little frustrated by our inability to articulate that experience fully to anyone but ourselves. Oftentimes we cannot even articulate our own stories clearly to ourselves.[1]

From birth we move out into a different world. Our place is not directly attached to our mother's body, but now we are attached to a different body. Our lungs become a new umbilical cord. Just as that physical conduit once served as a passage of life with our mothers, breathing offers us a passage of relationship with all living beings in the world. We take breath into our bodies and exhale a changed version back out again. In this way we continually connect, in a less visible but no less viable way, to the physical body of our world. We are never nowhere, and for as long as we are alive we are bonded with a world where we experience movement, breath, and nourishment. We are here. To live is to be in relation to the body of our environment.

Because we are a places, we are locations, literally continually evolving points of view. No one but we have traveled our routes. No one else has seen out of our eyes or felt the joys, pains, and sorrows that make up our internal landscapes. Thoreau knew this when he asked in *Walden*, "Could a greater miracle take place than for us to look through each other's eyes for an instant?"[2] If we could, we would be able to see each other's points of view. Because we cannot, because our journeys are our own, we must find other ways to communicate what we know and have seen. We create art. We tell stories. We make love. We dance. And still we grope to explain the deepest parts of us and then to find others who embrace us and make us feel as safe, as connected, as we did when our mothers' heartbeats sounded like cathedral bells inside our own worlds.

But home is more than just a body, a place; it comes to mean very quickly that needs are met, voices are heard, desires fulfilled, and from that sense of safety new terrain can be explored. Exploration occurs only when it is rooted in safety. The borders of what define that sense of safety shift and grow with each new experience, allowing movement farther and farther away from the physical environment of home without losing the emotional safety it affords.

The radiating arms of home allow for the creation of a lifelong process of weaving stories. Much like baskets that are grounded by cardinal points at their centers, over time these ribs widen with the addition of new directions, strands, and colors, creating an interwoven complexity that can no longer be unwoven. As we expand experience through exploration we story our world, and those with whom we begin to share new stories become deeply embedded into our sense of identity.

These inhabitants of our first in-groups share first times swimming and riding bikes, first times spelling names in the snow, first times holding puppies or burning hands on hot stoves. The people we come to know in our community, the children with whom we play and dream, become fused into the creative imagination of childhood and later form the bedrock of the memory of home. We become the sacred keepers of each other's "what ifs." The summer days shared daydreaming, the building of December snow forts, and the leaping in piles of autumn leaves write themselves into our unconscious sense of home, and we taste them with nostalgia for having once inhabited a time of dreaming. We love those in our in-group because we have dreamed the world into being with them and shared parts of the sensuously intimate space of imagination. Together we have given birth to dreams of who we are and who we will be. It is those dreams that we try to protect, building walls to keep out those who would destroy them. People die defending homelands, protecting the touchstones of memory, but more preserving the landscape of their dreams of becoming. Our newborn dreams we protect with the same alchemy of love and fear we afford babies because they are our creations, and our visions of who we want to be are more important to us than who we are now.

Bachelard's vision of the childhood home as the seat of reveries resonates with this image of home. For the individual, home "shelters daydreaming, the house protects the dreamer, the house allows one to dream in peace."[3] When a landscape becomes filled with the collective dreams of a group of people, it develops walls bound by the power of a culture's shared myth of its own becoming. When threatened, leaders reinforce those walls with offices of "Homeland Security" and send those they believe do not share the stories of the people to the "Home Office" to see if they are going to be permitted to enter or belong.

Belonging means knowing a place as it has come into being and feeling like a participant in its process of change. Home is a place where we believe we have power to be agents of change capable of turning internal imagination into reality. Roger Hart records in *Children's Experience of Place* that children experience place both by being in the place and by manipulating the environment in a way that is pleasing to them. "Clearly," he writes, "the most important aspect of all building and landscape modification is the satisfaction children find in the process of transforming the physical world."[4]

Only at home can the environment be sculpted to reflect the curves and colors of our own imaginations. *Manipulation* comes from the Latin word *manus,* meaning "hand." Hand and mouth—we shape the home place by bringing forth the dream of who we are and then forging an environment where outer and inner reflect each other in harmony. Wood floors, open spaces, light. Sound of voices, music, words. We feel at home when home reflects outwardly who we are and what we dream. What happens to those who are deprived of the opportunity to shape an external environment?

According to Sophie Watson, who studied the impact of homelessness on women, a number of deep identity-related issues emerge. Watson formed her central research question around the question "If a woman's place is in the home, where is she without it?" Interviews with homeless women led to their expression of frustration in terms of the inability to shape environment and without being able to shape environment being unable to shape social encounters or the feeding of self. Her informants told her that "homelessness is no control over the decoration and furniture" and "nowhere to rest yourself privately. It affects people, not seeing your friends, not being able to do your own thing and cook your own food." Most poignant, the dissolution of identity came because "homelessness is feeling nameless. I feel I don't exist, I'm just a thing."[5]

At home our voices are heard. If we are not heard, as Watson's informants shared, we may feel we do not exist. Storytelling requires an audience, a community. Without an audience, without a place to tell those stories, the stories die or build up like a dam inside the individual who is looking for a place to put them. There is no more profound aloneness than the sense that no one hears or sees you. If there is no place to dream, and no one to share those dreams, how can we feel at home in the world or be a part of a community?

During the late 1990s and early 2000s, a woman I knew led a writing workshop at a homeless shelter. Many people have questioned why she would lead a writing workshop for the homeless. Communities seem interested in the development of jobs programs and making people "productive members of society." Encouraging people to write and offering a place where they can share their writing week after week doesn't increase their "productivity," but it has had a powerful impact on helping the members of her community feel "homed." My friend found that sharing creative work, in much the same way as we do in our childhoods when we share our dreams, creates a community and from that community comes a sense of strength because voices are heard.

As many homeless people who call out on the street in hopes of an answer know, to be heard acknowledges existence. If you can hear yourself, and others can hear you, you exist. In the wild we may look to lose our language so that we can find ourselves. Without a home, we look to find our language and have others hear it so we can find ourselves. The making of place for many homeless people begins with the reclamation of voice.

Our voices are heard at home. We speak a shared language and one filled with pregnant references to other stories and histories that can be evoked with the simplicity of a gesture or a word. Our most complex language, the one we have the hardest time explaining to anyone else, arises from the storied landscape of home. When a language dies, so, too, dies a storied way of living with a place. The last speakers of a language feel the wrenching pain of never being able to be completely at home in the world because "you are what you speak." As Marie Smith, the last Alaskan Eyak speaker, records, "I don't know why it's me, why I'm the one. I'll tell you, it hurts. It really hurts."[6]

As children, my brother Gordon and I developed an elaborate play world involving our pet hamsters and a collection of stuffed animal toys that all became inhabitants of our imaginary village "New Hamsterdamn." Each animal had a name, voice, personality, and political inclinations toward one of the different hamsters and its ideology. With over one hundred animals involved, it was a complex world to outsiders but one we inhabited easily, developing ever-evolving stories of the political machinations of owls who rigged elections against elephants or the gossipy lives of hippos who spoke with strange speech defects and a pair of wild-maned twin lions named Leo and Zero who ran hair-styling salons. We not only inhabited this world with our animals, but we developed our own language, named for the most demanding and obstreperous hamster we ever had. Doodles McGurk. Her language was called Doodlish, and I was a fluent speaker of Doodlish when I spoke in the voices of the animals who were her followers.

As we grew older and the hamsters died off, we played less and less until eventually the stuffed animals were put away in the attic and only a few favorites kept as childhood reminders. Although we no longer play with those animals, the remnants of that civilization survive in an encoded language that Gordon and I still use with each other today. While we rarely speak it, our childhood lives in the language of that era, and as the lone speakers of Doodlish we inhabit a deeper sense of our historical home through the invocation of its language.

Leslie Marmon Silko echoes this sentiment in her experiences with the Pueblo community. She writes that "language *is* story" and that an essential element of maintaining the in-group element of a community is that the language itself is evocative of a shared collection of stories. "At Laguna Pueblo, for example, many individual words have their own stories. So when one is telling a story and one is using words to tell the story, each word that one is speaking has a story of its own, too."[7]

Keith Basso shares this power of place-names among the Cibecue Apache as a way in which language serves as a touchstone for journey and imagination. Here, too, the individual names and words serve as stories within themselves. The speaking of those names allows for the reverberating retelling of each of those stories. He describes the act of "speaking with place names" as the highest form of expression, noting that very short utterances, "like polished crystals refracting light," can capture compacted multiplicities of meaning for those who know the intimacies of the language.[8]

This intimacy of language that establishes connection to kin group is not limited to humans. Orcas off the coast of

British Columbia belong to a series of pods that are distinguished both by family group and by language. Three groups referred to as "clan groups" are distinguished by their calls. Thus clan A calls differently from clans G and R. Similarly, on the southeastern coast of Newfoundland over a million gannets nest on a rock outcropping beneath the lighthouse of Cape St. Mary's. Within moments of hatching, mother and chick know each other's calls well enough to distinguish their voices from the cacophony of a million other voices.

Homes are crafted not only by walls and windows but by voice. To be at home means that we can speak fluently but also that we are heard. Just as the orcas call back to each other and the gannet chick answers the mother's call, to feel at home we must be in a place where we can be heard and understood.

While the physicality of the space, the way in which it is decorated, or where the garbage bin is placed, all reflect aspects of the way in which people co-create a sense of home, it is the expression of voice that feels most elemental. In that, the inscape reflects onto the exscape the dreams and desires of a person or people so that the space balances harmoniously with their feelings of safety, security, beauty, and light. When entering into another's home we become immediately aware of the underbelly of his or her imagination—we know more about who they are simply by knowing how they choose to live.

For many, though, this issue of voice breeds the seeds of ambivalence about home as their voices are lost or shut out by the very people who brought them into the world. To have a voice is to have a presence, a power. In the alchemy of love and fear there are those who spend too much time living in fear of the voices of others, and those people may find themselves surrounded by company and kin yet still very much alone. So often people speak of an estrangement from the places they thought would be home because they are not allowed to "be who they are" there. It is hardly surprising that teenagers redecorate their rooms and resist attempts to have their place be dictated by adults during a time in life when they are deeply engaged in finding their own senses of self and their own unique voices.

Coming Home Again

After my parents died I experienced a sense of homelessness that was based on a sense that my voice could not be heard. It took a long period of quiet on my part before I was ready to try to speak again. In part, I believed there was no one who would know me with the same familiarity and intimacy I had known in my friendship with my father. I had to build a new sense of home.

For me, I built a new sense of home by returning to the one in which I was raised and rebuilding it as a reflection of who I am instead of who I was or who my parents were. They loved dark colors, brown carpets, and deep blue walls. Curtains were closed, and yew trees grew to filter the light in front of the windows. In rebuilding their home, stripping off the paint, pulling up the carpet, pruning the trees, and sanding the floors for days on end, I found a new sense of home, one I had never known before. In time, I even found a partner who could hear me, love me, and share my past history and a new one not yet written. He, too, sands the floors and paints his colors into the walls. When his father, Jim, cut himself working in one of the rooms and bled on the new white walls, I knew that his story would be written into our home, too. Jim has planted gardens for me in both of my homes, so he is always with us, even when far away.

Although Bachelard suggests that our childhood homes live in a place of "time immemorial," in living in places that hold the continuum of time we find strength in the complexity, the diversity, and the wonder that springs from knowing that a space can be in the same moment the keeper of a memory, the trigger of feelings, and pregnant with a new set of possibilities. If we are to become like Thoreau's saunterer, who is at home everywhere in the world, we must make our world a place where our hands do not destroy so much as shape and in our inscapes find a way to dream, love, celebrate, give voice, and listen to the members of our greater family, the myriad beings with whom we share our world. Perhaps once we are at home with ourselves we will be able to be at home in the world, too, for home is not only where the heart is, but in our hearts we find our deepest sense of place, too.

SIX

Lost Stories, Lived Places
The Lure of Ruins

> Notions of the past and future are essentially notions of
> the present. In the same way an idea of one's ancestry
> and posterity is really an idea of the self.
>
> N. SCOTT MOMADAY, *The Names*

Three images.

First image.

A July morning at the lower edge of Lake Nasser in Egypt. The sun rises, a ball of pure redness in a hazy white sky. Before the sun warms the sand, we disembark from our boat and walk across now cool sand to stand in a courtyard before the temple of Abu Simbel. Ramses II's stone eyes look across the flooded land of Nubia. A fictional story of his conquest of the Nubians and Syrians lines the entranceway to his shrine. The stones are original, but the place from which the stone eyes of Ramses gazes are different. To keep Abu Simbel from being flooded by the Aswn dam it was moved over the course of nine years, stone by stone, to be reconstructed in a perfect replica of its own self. Three of the four faces of Ramses II gaze off into Nubia. The fourth lies face down in the sand at the long dead pharaoh's feet.

Is a ruin still a ruin if it has been rebuilt, not as it once was but rebuilt anew as a ruin?

Second image.

A May afternoon in Manhattan. I walk past a church lined with snow-beaten teddy bears looking battered and forlorn. Thousands of origami cranes weave round a black metal gate, hanging between signed T-shirts, photographs wrapped in plastic sandwich bags, and a child's drawing of two buildings reaching out to hug each other beneath a blue crayoned sky. The spontaneous shrine, with its regularly rotating offerings, commemorates the story of a moment of instantaneous ruin. In one clear, blue September morning, the world watched in disbelief as two of its tallest buildings and those people trapped inside collapsed into a colossal ruin. For months afterward, people worked tirelessly at the sifting of metal, bone, stone, flesh, and paper. At night, two long beams of green light filled the familiar space with a ghostly shadow light. In time, construction will begin on new buildings, but, unlike Abu Simbel, the ruin itself will not be re-created. New life will fill the space of old. In City Hall planners argue over words like *memorial* and *marketability*.

Will people want to live and work in clear sight of the ragged eyes of teddy bear memorials and photos of the dead wrapped in small plastic bags? What will happen if they choose to forget?

Third image.

My immigrant grandparents bought their yellow stucco house in Brooklyn in 1924. For sixty-five years my grandfather tended his lawn, played chess on the front porch and Chopin on his piano in the front parlor. My grandmother made her clear chicken soup from chicken feet purchased from Saul the butcher, planted red geraniums in long white pots along the steep brick steps, and fed her children, grandchildren, and most of the neighborhood beef brisket and pickled herring. Twelve years after both were buried, I go to see their house, to touch the walls I knew in childhood. I want to know something of my mother by knowing the place where she was born and raised. I want to see the sidewalk where she fell off her bike and broke her two front teeth, the bed with the white covers where she wrote her first journals, the walls that housed her first dreams.

Chunks of stucco lie upended in the grass like beached whales. The driveway is filled with weeds. Ivy drapes the backyard in a parody of my grandmother's laundry line. The ivy pulls away the stucco, and yet its long sinuous arms give the little yard the feel of a cool grotto, like Aphrodite's pool in Cyprus, instead of the draconian order I once knew under my grandfather's rule.

I look through the cloudy windows and see through time to the house as I remember it. The house is still filled with my grandparents' possessions. Rusted padlocks block each door, but in time the ivy will pull away enough of the wall to create new doorways. One blustery fall day all of the pictures, letters, boxes of books, long unworn shoes, and forgotten toys will simply blow away.

My aunt cannot bear to empty the house nor can she sell it. She is tacitly letting the house fall into ruin. It has become the physical embodiment of her losses, a set of doors she cannot open but ones she will not let go. When I ask her for the keys to the locks she tells me, "Some doors should never be opened again." How quickly a home passes from lived to ruins. With such speed work the hands of wind, rain, and ivy. They may undo the constructions built by hand, but they have no power to erode memory.

Why would someone choose to turn a lived space into ruins?

Imagine soft afternoon clouds floating through the absent stained glass windows of Tintern Abbey or an abandoned fishing shack on the west coast of Newfoundland where salt water and time have weathered the walls to the softness of driftwood. Picture a dust bowl era tractor in a field in West Virginia quietly, gracefully, turning a perfect shade of autumn rust, as a ruin is, by its nature, a deeply complex visual embodiment of the deeply storied layers of time.

The Romance of the Ruined

Why do we love ruins? What is it that draws us to Hadrian's Wall, the Parthenon, Angkor Wat?

Like the wild, ruins are largely peopleless places. Unlike the wild, though, ruins are the remnants of attempts by people to manipulate their environment. The quality of peoplelessness is essential for sparking the imagination as ruins offer the perfection of the unfinished story. That story allows for the creation of two streams for the imagination. One stream follows the questions left in the trail of answering the question "What life did people lead when this place was at its prime?" We look at the type of buildings and structures, the broken pots and children's toys, and our speculations ask us to develop a story of who they were based on what bits of evidence they accidentally left behind. Who they are to us is a story of our own construction made of tantalizing bits of evidence from which we can receive very little verification. We interact with the ruins by imagining the story of the people when they were alive, a story that cannot be sullied by the verification of a live informant. We are free to imagine human life unencumbered by the need for specificity or accuracy. We enter into the arms of a true romance not in the sense of amour but in the sense of romanticizing.

The second stream follows the course of questions posed by time and human nature. "What happened to these people so that their place became ruins?" This the Ozymandias moment when we connect past to present and ask what happened to this people, this family, this civilization. This is the moment of standing in my grandparents' driveway looking at the ruins of my mother's life. Percy Shelley's words echo in their confrontation of the question of the hubris of history and the passage of time.

I met a traveler from an antique land
Who said: Two vast and trunkless legs of stone
Stand in the desert. Near them, on the sand,
Half sunk, a shattered visage lies, whose frown,
And wrinkled lip, and sneer of cold command,
Tell that its sculptor well those passions read,
Which yet survive, stamped on these lifeless things,
The hand that mocked them, and the heart that
    fed,
And on the pedestal these words appear:
"My name is Ozymandias, King of Kings:
Look upon my works, ye Mighty, and despair!"
Nothing beside remains. Round the decay
Of that colossal wreck, boundless and bare
The lone and level sands stretch far away.[1]

Is their disappearance a good lesson in history or a warning or both? We can look at the ruins of a Roman amphitheater and imagine the scope of the Roman Empire and see that it was not impervious to change. Luxor's empty temples and obelisks punctuating the blue Egyptian sky remind poets and pilgrims that empires come and go. In Chaco Canyon, Stonehenge, and Avebury we can touch stones placed hand by people whose names are lost to time and ask who were these people and where did they go? In my family's story, at least I know. But in time no one else will. Their stories will blow away. Mine will, too.

In a place once dominated by humans the confrontation of their absence haunts us. Unlike our home places, which are filled with our own stories, ruins are the places of lost stories. We know that they were once storied—we can read the graffiti on the walls of Pompeii—but we cannot know their stories. Instead, like our first encounters with the wild, where our lack of language for differentiation becomes the seeds of the creation of a new, relationship-based language, so the lack of story in ruins becomes the inspiration to imagine anew.

A place dying is a source of pain and shame. We do not want to be present for the process of falling into ruin. Unlike the World Trade Center's collapse, ruination is rarely a rapid process. We do not want to see the life go out of a place, feel the motion move from hopeful to regret. Ruins free us from the human experience of decay because by the time a place has become a ruin it has been filled with a different kind of life: ivy and moss, fox hollows and swallow's nests, blackberries and wild rhubarb. Although they are human ruins, they are also landscapes of reclamation, transformation, and rebirth.

Until recently Coney Island always felt to me like one of those ruins in progress. It was a short walk from my grandparents' home, and I was raised on stories of Coney Island as it "used to be" instead of as it was. In the late 1880s, Coney Island embodied all the extremes of its age. Opulence, graft, a giant hotel in the shape of an elephant with thirty-four guest rooms in its head, Coney Island was home to the first roller coasters, the first luxurious New York beach. The Coney Island subway stop, the very end of the F train line, still had fancy writing in its marquee, although many of the letters had fallen off over time. The diner in the entranceway felt like a perpetual Hopper painting where no matter what day or night it was always 3:00 am. No sunlight from the beach ever penetrated its terminal darkness, making it a depressive's mecca.

For decades, the burned-out ruins of the Thunderbolt roller coaster lay at the center of the horizon. Weakly protected from vandalism by a chain-link fence, it remained a memory of past dreams. Along its blackened curves grew lush green vines that took the shape of the coaster's track. Standing with fingers entwined in the chain-link fence, the cycle of death and life seemed so clear. As the roller coaster died, so life had grown over it. In the last few years the ruins of the Thunderbolt were finally taken away, and an empty lot now stands in their place. White-faced bindweed flowers twine their way up the fence that once kept us from the old roller coaster. The ground beneath it is littered with broken glass and Nathan's cups, but flowers rise from that same soil.

I can romanticize the ruins and the way vines intertwine with the roller coaster until I hear the cries of a child in the boardwalk shower. I see broken glass around her bare feet while her mother yells at her. Toothless men line the sidewalk of Surf Avenue selling the contents of houses like my grandparents'. Old toys mix with packages of pantyhose on blankets laid out for sale. They all smell of rain and mold. Some of the shoes were likely my grandmother's sturdy white plastic sandals.

Alone, devoid of toothless salesmen and screaming children, I can imagine Coney Island in its heyday, but with company, and barkers trying to lure people to the freak shows that still line the boardwalk, I cannot invent that story. I can romanticize the roller coaster before it burned, before tall grass grew up around it and bindweed took root. I can see it as it once was except that there are people there who are still trying to keep it alive. It is the same feeling I get in visiting nursing homes where I see the story of a past glory in those unsure as to why their present is a less vibrant version than their past.

Ruins in progress personify the failure of the myth of progress. These are the landscape of the last set of stories or dreams turned nightmares. The ruin in the making shames us all for not being able to maintain something at its glory. Once the people are gone, we are free again to imagine the place in its glory instead of bearing witness to its decline. We can imagine reclamation. Rebirth.

Earlier I suggested that one of the most important purposes of our encounter with wilderness is that it gives us direct access to the cycle of creation, change, and re-creation. Ruins serve a similar purpose only in ruins we experience the creative potential of death as the decay of one form leads to a different life growing out of it. In ancient rainforests these are seeds rooting in the backs of fallen trees, nursed by the very bodies of their ancestors. When seeds fall in the cracks of brick buildings and find root, it is no different. Nineteenth-century buildings in once vibrant neighborhoods in Cincinnati, Patterson, and Newark have become home to new forests growing on roofs and walls. In neighborhoods devoid of parks or greenery, trees grow from the tops of buildings, slowly deconstructing old buildings and factories in their own way. The farther their roots spread, the more the brickwork moves. While they are rarely permitted to create full forests in urban spaces, they speak to the tenacious power of life, a different life but one full of green leaves and strong trunks in places that have been left by humans to decay. George Simmel in his essay "The Ruin" suggests that this is an example of the power of nature over human hubris. "This shift becomes a cosmic tragedy," he writes, "so we feel, make every ruin an object infused with our nostalgia, for now the decay appears as nature's revenge for the spirit's living violated it by making a form in its own image."[2]

While I disagree with Simmel's interpretation and see ruins as an opportunity to see the power of the tenacity of life instead of an act of revenge by the natural world, I agree that ruins become filled with a nostalgia for what once was and will never be again. Like Bachelard's depiction of our first homes, ruins take on the patina of having once had a golden age in "time immemorial."

Ruins afford the opportunity to confront time, natural cycles, and the myriad ways of measuring time. To measure time in moss and lichen, or measure the time it takes for ivy to tear down a building or a small tree to grow in the cracks between two bricks, allows us to feel the transience of human "control" over nature and also, perhaps, gives hope to the idea that the natural world functions on a longer time scale. Wind, water, tree roots, and time can break down our houses, our roads, our attempts to bring our world back down to its elements. The Gaia hypothesis suggests that the earth will be able to heal itself if given enough time. Rebirth is possible but only if the relationship of humans working in opposition to the nonhuman world changes.

In ruins we have the place with no story. If the stories that come out of wild places often have the format of human conquering the environment, the stories of ruins are the opposite. In these, the environment is the victor over the attempts by humans to manipulate. Ruins foster the internalization of the feeling that creation comes out of destruction. Confronted with ruins, we create stories and, as Wordsworth once did at Tintern Abbey, "recollect our emotions in tranquility." Faced with the evocative power of juxtaposition, we are forced to confront our understandings of history, permanence, nostalgia, and time. When we believe our ancestry is connected to the people of those ruins, the connection is even stronger. To hold the broken chunks of stucco from my grandparents' house allows me to crystallize my grief into something I can hold in my hands and something, ultimately, of which I can let go.

Because ruins give physicality to the concept of life emerging from death, many ascribe sacredness to them. In the absence of live people to verify, this desire for a holiness of place ties into another complex set of stories to which our culture subscribes. Perhaps nowhere does our fascination with the coalescing of ruin, imagination, story, and speculation of the sacred come into play than in the explanations and explorations of prehistory. Prehistory allows us to merge our fascination with the beginnings of humanity as a part of ourselves and the tantalizing nature of having such little evidence left behind in the form of ruins. When the two come together, they forge a passion for the prehistoric, which may be fueled by the desire to find an authenticity in humanity and humanity's relationship with the natural world in the same way that so many desire to find the concept of an authentic home. That search is never an easy one and never one that is simple or pure.

The Passion for the Prehistoric

I confess, I am guilty. I work in a cave. In fact, I work in two caves. I try to understand what marks Paleolithic people left on the walls of two caves in France. I want to know who the individuals were who spent time in that cave. I don't believe that their markings are a sacred message from the past sent into the present. In fact, I do not believe their lines are sacred any more than I believe the words I write are sacred—which is to say that of course they are sacred and of course they are not. They, like my words, are part of a continuum of people telling their stories and living a life that creates new stories.

While many people are familiar with cultural ruins, often living in cities, such as Rome or Athens, that are part modern and part antique, not all people have had the opportunity to spend time in caves which were used in the Paleolithic.

To enter the cave of Rouffignac is easy. Moving from light to darkness seems natural when we work in autumn and unnatural on spring days when we would rather be out in the sun. Rouffignac is a large cave, over five kilometers deep. Today the entranceway is long and smooth, although it was once a lunarlike surface pitted with craters made by bears in search of winter refuge. The main chamber, once a riverbed cutting through the limestone of an even more ancient ocean, follows an even trail, rising only a little before turning a steep corner at a place where water drips from the ceiling and the air smells like wet cement. From there, the pathway slopes quickly downward, moving beneath a heavy ceiling and the remnants of past rockfalls. Lower still, the undisturbed floor undulates with waves created by cave bears that for fifty millennia made winter dens. Bears shaped the environment of the floor of the cave, but as far as we know humans did not. Instead, they drew mammoths and bison, horses, wooly rhinoceroses, and a lone bear on ceilings and walls.

We do not study the figurative drawings so much as we look at the lines made by human hands on walls and ceilings. Some intersect and overlap pictures of mammoths, while others stand alone, covering ceilings with long moving lines, zigzags, and parallel uprights. Panels emerge from the darkness stretching twenty feet with upright strokes of thin-fingered flutings. We call all of these lines Sevérines.

Two of the four chambers where we work branch off

from the main passageway. To reach the first requires a long sinuous route, our feet following the path of an old streambed now dry, round large fallen stones and small piles of dislodged flint. At first, there are long, flat, white rocks, but then there is the sensation of tumbling over smaller stones, turning, flowing over and around, down and down until coming to the chamber where the ceiling meets the wall, like the top and bottom of a scallop shell held together by a thin band of strong white muscle.

At the entrance on the left is a long set of vertical lines, and then the lines move along the ceiling, curling, waving, washing in long streams around rocks of flint. The hands upon measure are tiny—28 millimeters across—delicate fingers pressed together at first and then wandering over the arced expanse of the ceiling. White lines across red walls.

We measure, hoping to begin to recognize the width of a set of hands. I know them as 22, 28, and 34, finger widths moving with their own distinct imprint. In the spaces where I strain to reach, the fingers are broader, adult-sized hands of 38 millimeters or more. Across the ceiling we find evidence of children and adults together, deep in the chamber. In the low crawl spaces that children could easily have marked, the ceilings are clear, devoid of lines.

In the past, this chamber has been called the Serpent Ceiling or Macaroni Room as different theorists have tried to give the moving lines an interpretation from either religious iconography or simply the idea that one could see the lines as strands of spaghetti thrown onto the ceiling. One theorist believed that the ceiling represented a battle of snakes. Another has called it the "intestines of the underworld." We call them Mirian lines, a name that has no significance other than being derived from my elder stepdaughter's name, and the chamber simply A1. Too much of the vocabulary of prehistoric archaeology is laden with interpretation. The lines we study cannot be serpentines, macaroni, meanders, or water signs without having interpretation preloaded into them.

Farther into the cave, in a chamber known as E, a different type of finger fluting appears on the walls. This small red chamber has some of the most unique markings in all of Rouffignac. In the circle of an inverted cauldron on the ceiling, are long single-digit marks made by an adult-sized finger. Sets of sixes and sevens, the line patterns repeat over the span of the ceiling. In some the white lines have later been retouched with clay, different clays, as some are brown and others red. We find the remnants of charcoal, a burnt stick, in the weaving pattern of one of the sets of markings. I cannot reach these marks at my height, but Kevin, standing over six feet tall, can, and I see him in the shadow light, holding up a finger to imagine the feel of the maker, moving a lone finger down the slope of the ceiling.

On my knees, trying to avoid the remnants of an ancient charcoal fire pit, I measure Mirian lines. These long lines look radically different from the Kirian lines, a type of line drawn with a single finger that Kevin examines above. Small hands again, some at 34 millimeters, some smaller, running the length of the low ceiling. In one spot, where I can stand, there is a pair of hands that have started across from each other, then crossed, as if the maker were conducting an orchestra and we have the markings those hands made in the air in front.

"It is an accident that we have these paintings and markings at all," our guide, Sévérine, comments as we pause below the markings. "The people who made these did not intend them for us. They had no sense of time."

"Do we?" I ask her, still contorting myself with my ruler beneath the lines.

"No. I can imagine my children, and my grandchildren, and perhaps a hundred years, but after that, no more."

She is correct, and yet when I found a hand that matched the width and profile of mine, with at least twelve thousand and more likely twenty thousand years between us, I could not say that we were genetic kin, family, or anything more than two people with a 38-millimeter hand span. Yet the parallel of my hand to that hand drove home the feeling of being connected. Knowing this person, who did not know me, I am still aware that we have something in common.

Do we have a sense of time? We who cannot imagine what people two generations from us might know, can we imagine what those living twelve thousand years from now might think of us?

"They will study our graffiti in the metro" Sevérine laughs, an idea I have considered many times. "Will they think it is so important? Will they think we thought it was holy?" David McCaulay's lighthearted mockery of the tomb of Tutankhamen excavation, *Motel of the Mysteries,* drives home this point. In a futuristic world, a group of anthropologists discover the ruins of a pay-by-the-hour motel in Las Vegas at the heart of the lost culture of Usa. Their copious notes on the speculative meaning of their artifacts lead them to believe that the toilet was the seat of holiness. The author goes to great pains to show the wife of the excavating archaeologist wearing the holy toilet seat (with sanitized strip still attached) to the museum opening of the exhibit of their artifacts. With humor Macaulay reminds us how little ruins tell of meaning but how much we choose to assign meanings based on our own beliefs.

Many ascribe a sacredness to the caves out of a sense of amazement for the things they have survived over such a long period of time. The drawings are beautiful, the line markings varied in form, speaking to a multiplicity of reasons for their existence, but, looking at the way in which the Paleolithic is used by our current culture, the myths of pastoralism and the belief in the spiritual supremacy of "the noble savage" seem all too alive.

These caves, and the rock shelters in the region nearby, are ruins in that we find the enigmatic remnants of human culture without the story to support them. Unlike many ruins where humans have attempted to build something to "stand the test of time," our access to these spaces is perhaps accidental, as the climatic conditions of limestone caves allowed these human artifacts to survive. Again, we meet the questions of time and place, of process and the intersection between the nonhuman world and the human. Why, then, do the interpretations of the Paleolithic evoke a simultaneous sanctity and mockery? Is it simply that the farther back in time we explore the more extreme are the polarities of our narratives?

The Museum of Antiquities in Perigueux has a series of dioramas of prehistoric people: hairy men with well-situated loincloths and their topless mates sporting perky Barbie doll breasts. They are short, dark-skinned people. As with almost all natural history museums that have these displays, the men are engaged in killing, burying, or toolmaking. The topless women weave baskets, make pottery, and have a baby somewhere in tow. Everyone is always doing. Perhaps our capitalist society has difficulty trying to portray one that is not. A Calvinist work ethic imposed on hunter-gatherer societies fills our frozen models of their lives with continuous work. Anthropologists tell us that hunter-gatherers work much less than agriculturalists.

Museums have a terrible time finding ways to represent laughter, song, and relationship. The mundane beauties of individual life are difficult for the categorically reproducible nature museum display. If we are on the outside looking in, we want to look at "representatives" of a "culture" not at the foibles of the individual. Museums need to encapsulate something ideal, anonymous, yet utterly representative of the beliefs, practices, and behaviors of an entire culture. The American Museum of Natural History has, for the last fifty years, displayed its Eastern Woodland Indians collection with a gallery of nearly faceless mannequins.

Imagine instead something more realistic, perhaps a museum filled with "prehistoric bad jokes" or "embarrassing moments in hunting." Because earlier people are always portrayed as working hard, contemporary opinion has come to view these people as dour, always deeply engaged in ritual and a life of only practical applications. Chances are there was a fair amount of giggling from the ten year olds and copious eye rolling from impatient thirteen year olds. Is there any culture in which that isn't the case?

In the representation of prehistory there is almost never a consideration that women and children may have been the cave artists. There is no hard evidence to suggest that the cave artists were male. The evidence from Rouffignac suggests the opposite as the vast majority of identifiable hands are those of women and children. In magazine articles and books, however, representations of the cave artists always depict them as men, hairy men evoking Fred Flintstone or Ringo Starr.

When reading pieces like this one by David Lewis-Williams, a South African cave art specialist who has popularized the theory of shamans as cave artists, the desire to associate all cave art with ritual and seriousness seems striking. He writes in the preface to *The Mind in the Cave:*

> Anyone who has crouched and crawled underground along a narrow, absolutely dark passage for more than a kilometer, slid along mud banks and waded through dark

lakes and hidden rivers to be confronted, at the end of such a hazardous journey, by a painting of an extinct woolly mammoth or a powerful, hunched bison will never be quite the same again. Muddied and exhausted, the explorer will be gazing at the limitless terra incognita of the human mind.[3]

Lewis William's language transfers the terra incognita normally associated with the wild to the landscape of the human mind. Does that make these people as unfathomable as the wild or as "other" as the wild or both? Does he need an association with wildness if he is going to claim that the cave artists were all shamans who bridged the landscape between the sacred and the mundane? He is not alone in this association between Paleolithic peoples and the wild, as many contemporary nature writers engage the image of the sacredness of prehistory and cave art in an attempt to create a pure lineage of human wildness. Bill Kittredge, Paul Shepard, Annick Smith, Terry Tempest Williams, Gary Snyder, Andrew Schelling, and Jack Turner all draw from encounters with rock art or prehistory in their work. Why does nature writing lead to these connections to prehistory?

For many there is a sense of the purity of the indigenous wild, a romantic image of a world unsullied inhabited by wise noble savages who lived in respectful harmony with the animals around them. Prehistoric indigenous people become tantalizing because there is the desire to assume that they had greater wisdom about how to live in the world than we possess. On the great mythical escalator of time, the farther back in time the people the purer they are. Invoke the horses of Lascaux to go back fourteen thousand years, invoke the bulls of Altamira to go back twenty thousand, and invoke the glorious lion panel of Chauvet to go back thirty-two thousand. Add inaccessibility to ensure that these places are holy because they are seen by very few. Since the majority of recognizable art depicts animals there is an assumption that the people who were the artists had a deep relationship with the animals around them, a relationship that we cannot understand in our world today. Something old, something lost, something wild.

Prehistoric artwork becomes impregnated with meaning to those who want to create a wild ethic that they believe has its patrimony in the Paleolithic. This simplification of meaning undermines the very concepts at the root of wildness, instead still embracing the tragic tradition's approach to knowledge. The Paleolithic's narrative becomes a tragedy, something pure, wild, and lost. Andrew Schelling's *Wild Form and Savage Grammar* is a fine example of this notion. Locating himself "30,000 Years After Chauvet," Schelling believes there was a "nearly forgotten contract between humans and other life forms,"[4] which, through the fusion of art and ecological knowledge, could lead to a "wild form." The salvation of our current world lies in the traditions of a people about whom we know almost nothing. He sees their understanding of the natural world, evidenced solely in their artwork, as "a way out of the West's goofy pastoralism."[5]

Because we know relatively little about Paleolithic peoples, speculation leads to easy exploitation. Sadly, these sorts of writings encourage a simplified assumption that these people had one unified culture. According to such lumping of time, this culture spanned a forty-thousand-year period with relatively little difference so that the artists of Chauvet at thirty-two thousand years would have been well known to those who came along fourteen thousand years later, as if time did not function in the same manner for these ancient people as it does for us.

While I do not disagree with the need for a deeper understanding of ways-of-being in the world, and that there is much to be learned from all cultures in all times, the use of Paleolithic peoples as the root of the human-wild contract serves only to reduce those complex people to an unverifiable simplicity with little difference from the hairy models at the Museum of Antiquities in Perigueux. In one context, the savage wild people are meant to be in negative contrast to the technological developments of the modern world. In the agendas of writers such as Schelling or Jack Turner they serve the polar opposite purpose of representing a purity that is unattainable again because of our technological developments. Schelling asks plaintively, "So 30,000 years after Chauvet is such an art possible?"[6] Is this any different from the conservators at Abu Simbel reconstructing the great Egyptian monument as a ruin?

These yearnings of writers to connect to the purity of the wild spring of humanity do more to create the dualisms of the tragic tradition than they do to resolve them. Consider Turner's first essay in *The Abstract Wild* where he centers the narrative on his encounter with rock art in a remote area of the Canyonlands National Park in Utah. On his first visit he is mesmerized and later, when he returns and others have seen the same pieces he has seen, he complains of feeling that the experience has changed from one with special power to one in which "they were just things we were visiting. I had become a tourist to my own experience."[7] He ends the piece by romanticizing the people who made these drawings, writing:

> But I wish we were wise enough to preserve something more. I wish that children seven generations from now could wander into an unknown canyon and receive at dusk the energy captured by a now-forgotten but empowered people. I wish these children could endure their gaze and, if only for a moment, bask in the aura of their gift.[8]

The assumption again is that if only we knew what they left for us we might understand better how to live our lives. In the language of self-hatred our culture believes that we must mourn for something we have lost, and the fact that we cannot know what that lost thing is makes us yearn for it all the more. This romanticizing of the people of past cultures does not free them to be complex, varied people; it makes them sacred in a way that is simplistic, unidirectional, in their "now-forgotten" but "empowered" state.

Gary Snyder suggests, in his *The Practice of the Wild,* that the greatest earth religion of the past was that of the cave painters of the Pyrenees. He writes:

> It may be that in the remote past the most sacred spot in all of Europe was under the Pyrenees, where the great cave paintings are. I suspect they were part of a religion center thirty thousand years ago, where animals were "conceived" underground. Perhaps a dreaming place. Maybe a thought that the animals' secret hearts were thereby hidden under the earth, a way of keeping them from becoming extinct. But many species did become extinct, some even before the era of cave painting was over."[9]

The stories to which Snyder refers come from the same artistic speculation that has led those who have studied cave art to refer to animal pictures as hunting magic, totems, sexual symbols, and the work of trance-induced shamans. For the past hundred years, each generation has produced a new "interpretation" of the cave art, and each interpretation fills the imagination of the culture and corresponds to the mythos of the times. None of these interpretations can be verified, as we have no shamans, no long dead hunters, no cave artists with whom to speak. Thus the caves, like ruins, allow the romantic imaginations and political agendas of the age to be written on them, as if engraved in stone.

Nature writers such as Schelling, Turner, and Snyder want a heritage that connects them back to other storied people who value place and wildness. It would be best if those people had the same ethnic heritage since the Eurocentric are vaguely nervous about roots going too deep in North America. Better to connect with the ancestors in the cave whose bones are harder to dispute. Of course, thirty-two thousand years after Chauvet it is hard to say who has directly come from the line of the lion painters so they are free for all to claim. Jean Clottes, the former head of the Chauvet research team, has offered tacit permission for this. Schelling quotes him as saying that the work of the artists of Chauvet "is not simply regional or national but in fact belongs to all humanity."[10] Thus are all then free to trace their own lineage back to those people with impunity or to think of these prehistoric people as the wellspring of creativity in humanity? Terry Tempest Williams uses this mode of connecting her Mormon cultural traditions to those of the painters of Altamira in this passage from *Leap*.

> The closing words from the Prophet ring through-
> out the stadium: We have a
> divine mandate to carry the gospel to every nation,
> kindred, tongue and
> people. . . . We must grasp the torch and run the
> race.
> Grasp the torch. Carry the torch. Pass the torch,
> The torch I see is not the burning torch of truth,
> not the torches lit with the intent
> of martyring a prophet, but the small handheld
> torch carried into the cave at
> Altamira in northern Spain, a torch fuelled by ani-
> mal grease that illuminated the
> dark so an artist could paint a bison galloping on a
> ceiling of stone.[11]

Does Williams's leap from a modern day Mormon prayer meeting in a stadium to the cave painters of Altamira offer a deeper connection between people separated by twenty thousand years or a desire to authenticate her experience? Why must Paleolithic people painting in a cave represent the purest spirit of being human? Twenty thousand years from now, what evidence will we have left to be found by future archaeologists of our purest spirit and creative selves? Or are we too tainted in our times by technology to even imagine the possibility of such a discovery?

Brian Sykes, whose work in genetics traces all persons of European ancestry to the lines of seven women, offers people an opportunity to trace their genetic lineage back to one of the women he calls *The Seven Daughters of Eve*. He suggests that we all want to know our genetic heritage, to know to whom we are connected and that our DNA "is a traveler from an antique land who lives within us all."[12] Perhaps nature writers are looking for a lineage connecting humans from the past to the present that will allow them to write with the freedom to generalize. By connecting to Paleolithic peoples it is possible to make generalizations about the basis for current humanity. Without the distinctions of culture, researchers and writers can speak of the human desire to create, to paint, to write. The lack of specificity makes Paleolithic people the ultimate "Everyman," the ultimate parental "Eve," since none of them is currently alive to argue, to deny, or to take notice of the appropriation of bones or beliefs.

Does the use of the Paleolithic by writers interested in encouraging ecological views of the world help or does their use of them only strengthen dualistic myths about purity and ruins? For many, there is the belief that an ancient earth wisdom exists that can be unlocked by deciphering what was left behind on cave walls. Those who live with less technology and closer to the land are perceived to be more deeply spiritual than those who inhabit a technologically advanced society. Do we always feel a simultaneous tug that we have both lost something in a holistic understanding of the natural world and gained something sinister in our Faustian bargain for knowledge of technology? If their art were unattractive to us, their hands sloppy and their animals cartoonish, would we want to be connected to them at all? Ironically, there is little written by American nature writers about a passion for Australian rock art, though some of that dates as far back as sixty thousand years.

A tremendous outpouring of human self-hatred is tied up in this dichotomy where the feeling is that it was always simpler and better "in the old days." That may be true. The fantasy of a simpler, holier past has served all technological societies well by providing a powerful dualism so that people do not invest in the present to make it a positive reality but yearn for an unreachable past. When that cannot be accessed, people feel less shame in destroying an already sullied present. Our desire for holiness in the past and a cultural sense, once again, is a move toward anticomplexity. The idea, the idyll, is a simpler, more spiritually unified way of life. Simplification may make for easy consumption, but it strips everything bare and allows the creative life, movement, and messiness to be removed. Would the cave artists have wanted our worship? Will the graffiti artists of New York someday be exalted with the same sense of wonder? We cannot know what will last, nor perhaps did they, so the holiness we ascribe in our moment shows more about how we feel about time and its impact on place than it does about our real knowledge of learning another culture's belief system.

While I disagree with Snyder's use of myths of the purity of the people who painted Niaux and Bedhilac, his overall understanding of a lived practice of relationship with the world of the present speaks to an understanding of the broader sense of living in the present instead of dwelling in the past.

> The point is to make intimate contact with the real world, real self. Sacred refers to that which helps take us (not only human beings) out of our little selves into the whole mountains-and-rivers mandala universe. Inspiration, exaltation and insight do not end when one steps outside the doors of the church. The wilderness as a temple is only a beginning. One should not dwell in the specialness of the extraordinary experience nor hope to leave the political quag to enter a perpetual state of heightened insight.[13]

We love ruins partially because they have lasted and partially because they have fallen apart. The story, like slow-cooked meat from the bone, has separated from the story-teller, but the place has remained. When we come to ruins, we look for ourselves in the ivy-strewn walls. It doesn't work if there is a live interpreter reminding us that what we see there is not romantic at all; in fact it is either practical or challenging or at least something more deeply complex than what we would like it to be—the sign of a simpler, wiser time than our own now. In connecting prehistory to ruins we see that same desire played out on both person and place. We are of a storyless people—we want their best story to be ours, too.

Returning to the three images that began this chapter, are we closer to determining why we are drawn to rebuild the ruin of Abu Simbel looking out from two thousand years in the past but are simultaneously eager to see something reborn from the space in Lower Manhattan where the memories of people whose lives were cut short lie buried? On the broken front steps of my grandmother's house red geraniums and irises still bloom untended in the pots she planted long ago. Although she has been dead for twenty years now, her flowers continue to bloom. None of her new neighbors, who see her ruin of a home as only an eyesore in their otherwise revitalized neighborhood, know her story or the history of the roots of those summer blooms.

Ruins are powerful places because, like wilderness, they are catalysts for storying and the imagination yet are more evocative of our human nature and our relationship with time. In ruins we create a companionship to our aloneness, a company we have invented from artifact and one breathed to life through the heady power of imagination where the setting teases us to write the story. They allow us to encounter the process of decay and change, life from death and life into death, and they let us know, through their interaction with the nonhuman, what it means to be human and how we feel about "what we leave behind."

The beauty of ruins is in their ability to capture our thoughts on history, on preservation, on decay, on romance, on our selves in the simple encounter, perhaps, of a rainy day, the outline of a tilting wooden shack, and within it the dream of what once was, who we once were, and how much we feel they are like us as we all try to find our way home.

SEVEN

A Gifting World

My teacher once said to me.
Become one with the knot itself
'til it dissolves away.
—sweep the garden
any size.

GARY SNYDER, *The Gary Snyder Reader:*
Prose, Poetry, and Translations

Last year Kevin and I took his eighty-one-year-old mother to
Pelorus Sound, the remote region on South Island in New
Zealand where she was raised. Pam is a quiet and thoughtful
woman, someone who often spends her time anticipating
others' needs so she will be able to meet them with seamless
generosity. She asks little and takes even less from the world,
often speaking of a quiet sense of gratitude for the small
things. Blackbirds singing. Fresh strawberries and peas from
her garden. Both she and her son surviving cancer.

The one dirt road that serves as the high road through
Pelorus Sound is an exercise in balance as it curves sinuously
over two-thousand-foot cliffs with not so much as a guardrail
to tease the imagination into thinking there is some protection between car wheels and the turquoise waters below.
When Pam was a child in the 1930s in Garnes' Bay, there
were no roads into the Sounds. The bays were only accessible by boat, and travel from one bay to another, one house
to another, required long walks over steep mountains car-

peted with thick native bush. To hear Pam speak of it, she wasn't raised with any hardship or in the wilds. She saw herself in her very own paradise.

Pam's parents, transplanted socialites from Wellington scratching out a farmer's living during the Depression, were ill-suited to life at a home they called The Pines. Their daughter, on the other hand, befriended the very place itself, becoming a playmate to the trees, the birds, and the water. Seventy years later, as we drove Pam back to the country of her youth, her relationships were still as alive, as powerful, as they had been when she was a child.

"*Hello Cissy Bay . . .*" we'd hear from the backseat as the car crossed over a thin saddle of land between two bays. "*Good-bye Cissy Bay,*" she'd whisper as we drove farther along. Her past personal relationships with these bays had been at the pace of walking feet, not car wheels, but, nonetheless, in the week we spent in the Sounds, Pam did not miss an opportunity to greet a place that was familiar nor to say good-bye to it as she left it. She shared with us the changes they had seen in seventy years, the stories of the people she had known who had lived in each bay. In saying their names she released the stories that lay dormant like seeds waiting for spring rain to sprout anew.

Pam did not greet these places to inform us of their names or to formally pass on stories. We had a map. There were road signs and guidebooks. Instead she greeted these places because they were her oldest friends and companions, familiar landmarks of her internal landscape. Just as when she greets the blackbirds in her garden and whispers hello to her strawberry plants, Pam unconsciously engages in a practice that I believe we all do with places we love—she anthropomorphizes them—so that she is free to love and feel loved by the world that surrounds her.

Mum would be very surprised if I were to tell her that she engages in an ancient behavior held by anthropomorphizing people throughout the world or that her worldview ties her directly to the people whom her son and I study in the caves in France. She would probably simply tell me that she attends St. Mary's Anglican Church in New Plymouth every Wednesday morning, as she has for almost sixty years, and certainly wouldn't want me to think of her as my prehistoric mother-in-law. Pam would not say her spiritual beliefs are anthropomorphic, but I would argue that her root behavior is. Of course, as a good daughter-in-law, I know better than to bring it up. Instead I ask her to show me the blackbird chicks in the nest outside her door. She says "hello little ones" to them. They cheep back.

Stories and Belief

How do we go about developing systems of belief? Does my mother-in-law actively seek anthropomorphic relationships in a world in which talking to animals, much less whole bays and valleys, is considered cuckoo? Why would she pursue those relations when her culture's mores contradict her experience? Wouldn't she rather fit in with the neighbors?

Pam engages in an act that all humans do; she stories her world. Our complex brains are in large part organized around the creation of story and narrative.[1] As we take in endless amounts of data from all parts of our bodies in any given moment, it is the capacity to story that allows us to process those experiences into something coherent. Story is a flexible medium, drawing different data at different times. In some versions of the same day we remember the weather, in others, dealing with the same events, we focus on the people we met, while in others it is a complex web of information that we interweave into our understanding of events, linking one story to another in a continuous tapestry.

What seems clear is that our capacity to create storied relationships with the world allows us to store multifaceted information about social relations, history, place, time, and emotion. Stories are by their very nature meaning-seeking activities, offering us the opportunity to bridge gaps, forge connections, comprehend time and space, and feel a sense of resonant harmony. Belief systems are rooted in the notion of collective narratives in which a group of people believes in a common story. Belief is then strengthened by experiential verification of story (thus people who pray and have an experience in which the subject of their prayers appears to be answered feel their personal beliefs strengthened). For my mother-in-law, her belief in a communicative and relational world is verified every day in one small way or another. She calls to the blackbirds, and they answer her. Her attempts at communication are verified—she believes her blackbirds know and recognize her as their friend and that her roses appreciate that she is more than willing to wrap them in burlap coats against a winter's chill. Her experiences of spring roses blooming wildly let her know that her small efforts have been appreciated.

Anthropomorphizers and the In-Group

My mother-in-law is an anthropomorphizer living in a largely anthropocentric world. She has to disconnect her lived experience of the world from the social mores of her community if she is to live successfully among most other humans in her New Zealand society. (My father-in-law, luckily, is just as much an anthropomorphizer, often referring to his flats of spring seedlings as "the children" and even naming the new dishwashing machine Sophie.)

If my in-laws were to live in another culture, perhaps one in which anthropomorphism is explicitly valued, they might be able to feel less of a disconnect between their lived experience and their culture. Among the cultures in which they might feel most comfortable would be hunter-forager peoples who place anthropomorphizing or animism at the heart of their belief systems. Transplanted to living in one of these cultures, Pam and Jim would likely find that they live in a small human population group that is linked to others linguistically and/or through shared beliefs, that their community has a nonlinear view of time, and that form and shape are less important than essence. The dominant belief system of their new community would be one in which they live in an anthropomorphized monist world where the non-human world is as filled with intensions and intelligence as the human world. In their new community they would also note that death is perceived as being part of a cycle involving nourishing the life of the world instead of being focused on the endpoint of individual existence. Given their regular tending of the plants and trees surrounding their grandson's and parents' graves, they might find this piece of belief to be the most resonant of all.

My in-laws tend not to seek out expert opinions on topics such as the duality formed between anthropocentrism and anthropomorphism. (They simply talk to birds, bays, and dishwashers, leaving the polysyllabic words to their progeny.) If they did, though, they would find themselves at the heart of a series of debates on the ways in which the contemporary Western worldview contradicts the experience of animists. "The major difference between American Indian views of the physical world and Western science," writes the Lakota theologian Vine Deloria from the heart of this debate, "lies in the premise accepted by Indians and rejected by scientists: the world in which we live is alive."[2] Ethnohistorian Calvin Luther Martin further suggests that hunters in particular see themselves as "the narrators and symbolizers of the blueprint of creation."[3] Humanity's role as they describe it, is not to name and steward creation, like Adam in the Garden of Eden, but instead to use minds "capable of imaging the vast yet interconnected network of creation and rendering it in language and material structure" to serve as the world's "historian-regenerators and artist-regenerators."[4]

Stewart Guthrie, an anthropologist who has studied the relationship between anthropomorphism and religious belief, adds, "Little or nothing escapes being anthropomorphized at some time or place and at some level of thought. People see animals, plants, artifacts, inanimate phenomena such as wind and rain, and abstractions such as death and time as more or less humanlike."[5] He argues that anthropomorphism is a brain reflex that allows us to make snap decisions about whether or not something is dangerous because of its difference. Anthropomorphizing is *not* a desire to see dogs dressed as humans or to revel in Mickey Mouse's antics so much as it is a way of ascribing the highest predictive capabilities our brains possess to the nonhuman world.

Imagine if my father-in-law were out walking at dusk on his way home from trout fishing and he saw a large, round shape on a hill above him that seemed to be moving a little. His mind might offer him the option of the thinking of the shape as a rock or a bear. (Luckily, there are no bears in New Zealand, so if he thought he saw a bear we'd know he had been walking home from a detour to the bar! So let's imagine him on holiday in British Columbia instead.) If the shape were moving, his mind would likely err on the side of the shape being a bear, and once he thought about the bear he would immediately begin trying to predict what sort of behavior the bear might exhibit. Guthrie calls this the "better safe than sorry" approach. He offers the idea that

> We animate and anthropomorphize because, when we see something as alive or humanlike, we can take precautions. If we see it as alive we can, for example, stalk it or flee. If we see it as humanlike, we can try to establish a social relationship. If it turns out not to be alive or humanlike, we usually lose little by having thought that it was.[6]

Further, by establishing social relationships with nonhumans (a strategy that offers a long-term survival mechanism), we also benefit from feeling a sense of harmony in which my father-in-law could feel consistently a part of the community of life around him. Whether bears, trout, mosquitoes, or buddies at the bar, all would display knowable and somewhat predictable behaviors. What would matter most to Jim is what sort of continuous relationships he has with all of these creatures and how he fits into the greater community of the place. For my own father, who spent most of his life studying animal behavior, this capacity to understand and predict was one of his greatest sources of continual wonder about the world since there was always something new to learn.

This predictive element raises the most central aspect of the anthropomorphic worldview. For anthropomorphizers, humans and nonhumans are members of what sociologists would call the same "in-group." This in-group is united by a shared belief that all beings are creations with a life force or spirit running through them. The life force (creation, creativity, nature and its laws), experienced by all living things, is the base commonality that allows for entrance into the in-group. It is socially reinforced by the maintenance of relationships through respectful or right behaviors. Martin describes this as a belief that "Nature conserves me, not I it," wherein language and behavior continually focus on the idea of the "interpenetration of the human with the other-than-human person." He describes the songs the hunter sings to the game as "You and I wear the same covering and have the same mind and spiritual strength," emphasizing essential sameness instead of otherness.[7] The anthropologist Edmund Carpenter further echoes this sentiment based on his experiences living among Inuit and Yupik peoples, wherein "the lines between species and classes, even between man and animal, are lines of fusion, not fission, and nothing has a single, invariable shape . . . but rather a sense of being where each form contains multitudes."[8] Or as the Pueblo writer Leslie Marmon Silko shares, "The Hopi way cherishes the intangible: the riches realized from interaction and interrelationships with all beings above all else."[9]

Nourishment and Consumption

Let's return to my father-in-law and imagine that he did not in fact encounter a bear, but on his walk home he did encounter his own ferociously growling stomach. A bad day fishing on the Arrow River brings me (and him) back from the spiritual landscape and into the more mundane reality of wondering how, in an anthropomorphic world, one can possibly eat if each time we want to eat something we aren't just eating *something* but *someone*—and if we believe in a world of "all my relations" then we are contemplating a relation such as that. In an anthropocentric worldview this is resolved by having a humans-only in-group where only humans are perceived to be sentient (and oftentimes, sadly, only certain humans). Thus, those who fall outside that group can be commodified. In the anthropomorphized worldview, however, all beings are members of the in-group of creation and thus cannot be *used* or psychologically converted into tools and crops. How, then, can a hungry anthropomorphizer find a way to balance the set of right social relations with members of the group and still fill his or her belly?

The solution lies in the system of belief wherein the continuance of life, in all of its forms, is paramount to the individual's existence. For the anthropocentrist, the centrality of the human experience requires a continual maintenance of dualisms—self/other, human/other, life/death—as a way of forming boundaries for an increased number of in-groups where humans fundamentally differ from all others. By focusing primarily on the human experience, anthropocentrism elevates individuality to a paramount position where each individual negotiates his or her relations with the divine (often through mediators) and final judgments are individualized. Anthropocentric worldviews often create linear time frames where corporeal life ends and a disembodied spirit continues on, following the same pathway as the corporeal life, just in a new, nonearthly location. Key to this belief system is the idea that individuality does not end with the termination of life—individuality continues onward eternally until a moment of world finality. In most anthropocentric worldviews, the maintenance of selfhood remains the central tenet. For without it how would one know that he or she is the center of the world?

The benefit for the anthropocentrist is that eating in this life is largely guilt free (not calorie free but guilt free) because the survival of humans and individual humans takes precedence over the nonhuman world. Paul Shepard contrasts this with the beliefs of Arctic peoples who "speak of hunting and eating not animals but souls." Shepard maintains that "the ambiguity of 'eating souls' can never be entirely resolved because we are both the swallowers and the swallowed."[10] Perhaps this is one of the sources of disconnect for anthropocentrists as they must resolve the question of continually consuming life to preserve the continuation of their own lives. Still, compared to the anthropomorphizer, the killing of an animal or the harvesting of a field of corn can all be done with greater ease because humans can cast themselves as the husbands and stewards of animal resources, creatures who are lower than they are in the hierarchy of sentience, instead of being kin to or relations of all beings.

Ironically, needing to mitigate the landscape of relations to be able to eat affords anthropomorphizers a psychological freedom that the anthropocentrist does not experience. For the anthropomorphizer must ask permission from animal or plant for the taking of life. He or she is offered a gift from the relation that has at its root the implicit covenant that "when it is your time to give up your life for another, you will do so, too." Thus, the anthropomorphizer is engaged in a tacit agreement with his or her community to be nourished and ultimately to also nourish the world—to literally become worm's meat or pieces of mouse nests when the time comes.

The psychological nexus of anthropocentric life, on the other hand, requires the maintenance of individual life. Instead of receiving gifts from relations who offer their lives in this tacit agreement, the anthropocentrist possesses the right to harvest resources because of humanity's innate difference from all others. This does not put him or her into a long-term covenant with the resource but instead reinforces the idea of a "single use" of a life. Since humans do not enter into the system of relationship with food, it ceases to be a reciprocal nourishment relationship of the life of all and instead is seen as the end. For the individual to persist, he or she must avoid death.

As a by-product of the anthropocentric emphasis on the individual (and belief systems regarding individualistic relationships with deities), one's own desire to prevent death, or to behave correctly so to ensure a specific afterlife, creates a system whereby the warding off of death through the acquisition of food, life lengthening, or death-warding/afterlife-affirming knowledge underpins economic and social systems. The high social status afforded to both doctors and priests in contemporary society exemplify this belief.

Theorists propose that the historical shift from anthropomorphism to an anthropocentric worldview took place approximately ten thousand years ago at the time when agriculture began to take precedence over hunting and foraging. Martin proposes the idea that the shift came not from a change in technologies, or even a change in weather at the end of the Ice Age, so much as a shift in stories and beliefs focusing on fear of an unknown future and potential death. "I think the underlying motive for the abandonment of simple foraging was fear. As I brood over the metaphysics of the Neolithic," he writes, "I detect a language and artifice ridden by fear: fear of not enough food, fear of animal elusiveness and hostility, fear of our own deaths." He adds, "I see none of these elements of fear disclosed in the voluminous literature, both historical and ethnographic, on hunter-gatherers."[11]

The power of this fear likely destroyed a monist worldview and led to the dualistic boundaries set by anthropocentrists. This shift in ideology led to a world in which consumption of nonhumans was valued because it served the greater goal of warding off death. Dualisms and monisms cannot coexist as a dualism will by its very nature attempt to polarize aspects of monism, thus creating irreconcilable divisions. The anthropomorphizing monists would not be able to combat a consuming duality in which some beings are sentient and others are not. They likely had to give in to the rhetoric of fear to remain in their human communities.

Ultimately, the economy of a death-fearing culture became based on the acquisition of possessions (food and others) with the underpinning desire driven by the belief that by stockpiling these items one can be protected from the unknown potential dangers of the future. The protection of those stockpiles, however, created further and further boundaries, and access to those stockpiles led to more ingroups, power structures, and social hierarchies.

What, then, is the "higher good" of the anthropocentrist? It appears that feeding the economy is the way in which one actively engages in providing protection from death. As one acquires more, one believes in the possibility of warding off death for a longer time. Perhaps this is why, in the days following the September 11 events in New York and Washington, the U.S. president exhorted the country to "go shopping" as a means of aiding the psyche of a grieving country. Consumption of resources, acquisition of goods, and laboring toward the ability to engage in higher levels of consumption become an insurance against fear. For the death-fearing ideology, decisions must be reduced to that which can be consumed easily and time compressed to the immediate on the cusp of an unknown and ever-looming future. The "buy now—don't delay" approach to modern advertising has only reinforced an economy of fear and has contributed to the wild pace of contemporary human consumption of the nonhuman world.

A Gifting World

Let's return to my father-in-law and his growling stomach, for he is caught between two worlds. In large part, he and my mother-in-law grow many of their own fruits and vegetables from the seeds he nurtures. Still, they are happy to shop at the Pak 'n Save and, like good Kiwis, never miss the opportunity for a spot of roast lamb. Were my in-laws to live among their imaginary hunter-forager group of anthropomorphizers, how might they reconcile the feelings of kinship with nonhuman beings and also needing to eat?

The answer lies, as I shared before, in anthropomorphizers' deeper belief system. At the heart of this system rests the belief that they are participants in creation, *which is greater than themselves.* The goal of life is not for them to continue onward but for *life itself* to continue. Deloria describes this among Native American/First Nation peoples as follows.

> The task of the tribal religion, if such a religion can be said to have a task, is to determine the proper relationship that the people of the tribe must have with other living things and to develop the self-discipline within the tribal community so that man acts harmoniously with other creatures. . . . [T]he awareness of the meaning of life comes from observing how the various living things appear to mesh to provide a whole tapestry."[12]

One of each individual's roles in life is naturally to live, but to eat each person engages in a covenant. This covenant is formed in the moment of asking for another's life in the present because it simultaneously requires each of us to also serve as nourishment for another when the time is right for us to give up our lives. Deloria concludes that "for the tribal people, death fulfills their destiny, for as their bodies become dust once again they contribute to the ongoing life cycle of creation."[13]

The wealth of life is measured not in the acquisition of possessions but in the maintenance of relationships in a world of kinships where the life of one is interwoven or interconnected with the life of all others. The system of nourishment, then, is based not on the idea of consuming and stockpiling toward an unknown future but is instead rooted in the offering of gifts toward the continuity of life in the future.

The idea of gifts and thankfulness permeates the language and social mores of societies that anthropomorphize. They see gifts not only as the exchange of commodities but, more important, as a living spiritual force that allows creative energy to bind and continually revitalize the community. For hunters who take the lives of other animals, the way to mitigate the feelings of having taken another's life is to enter into a system of gift exchange whereby the animal has offered its life to the hunter as a gift and then the hunter must honor the gift and pass on his or her gifts to another. Because all participants are members of the same in-group of "creations," they can all equally engage in being gift givers and receivers. According to Lewis Hyde, what is essential is not the actual gifts themselves but the social binding power (reminding us, perhaps, of the etymological roots of the word *religion, religio,* meaning "to bind again") in the act of giving as the gifts contribute to the life of the whole. Gifting becomes the central ethos of the culture.

Hyde sees direct interconnectedness among creation, creativity, and gifts. In his seminal work, *The Gift,* he describes gifts as needing to possess four main qualities. First, the gift offered must be the fruit of one's labors, one's passions. In this way, the gift is an expression of the life force that has flowed through and nourished the individual. Second, the gift must be offered without belief in a sense of reciprocity—it is not a payment or an exchange but a simultaneous offering of full self given in complete selflessness. This, too, honors the life force by not attempting to possess it but by passing it on in its changed form. Third, the gift nourishes the giver in the act of giving it away as wealth is gained through relationships and relationships are strengthened through generosity. Finally, new gifts come to the giver when the giver has engaged in the act of giving away because the life force functions much like a river, continually flowing and filling that which has emptied. Hyde notes, "[T]he only essential is this: *the gift must always move*" because the gift itself is the spirit of life.[14]

How does this fit with my hungry father-in-law? His emphasis, if he truly wants to maintain a relationship with his nonhuman kin, must be in offering up his personal gifts. What are his gifts? If I asked him, he would say he's pretty good at repairing a clock since he'd spent a lifetime as a watchmaker. Chances are good his future trout dinner probably doesn't have much interest in a grandfather clock or even a really nice Rolex. What, then, are his truer gifts? He might have to look a little farther into himself and say that he tends to work at things until they're done. He can wait patiently and work with great persistence at what appear to be insurmountable tasks. He would say it brings him joy to nurture things and see new growth. Even more than that, he would share that he openly loves his wife and cares deeply for her happiness. He worries more for her than he does for himself.

These are more along the lines of his true gifts. In his new community of anthropomorphizers, his values and gifts would be well received. Martin suggests that those gifts of selfhood celebrated among hunter-foragers are largely about learning how to live well with others through a feeling of true generosity not commodified exchange. In his book *The Way of the Human Being*, he describes the Haida belief in the *xhaaidla,* a skintight membrane stretched tightly over the world.

Passing through the boundary one emerges into something else while retaining the essential nature of one's former self. What remains is the beauty and power of what we might call the Common Self: the deepest meaning of kinship, in other words. For, having penetrated the membrane, one now begins living from the vantage point of that other being who is, in the end, one's real self as well—one becomes the Common Self. To learn this is to become a truly genuine person.[15]

Existence for the anthropomorphizer is largely about helping to reveal the beauty of the world not as a static entity but one in dynamic and continuous creation. Participation in this role is through the creative gift. The First Nations poet Diane Burns describes this by saying that "creativity is the force that propels everything. That's the energy of the universe. The more you feed it, the more there is to go around."[16] Deloria adds:

Ceremonies have very little to do with individual or tribal prosperity. Their underlying theme is one of gratitude expressed by human beings on behalf of all forms of life. They act to complete and renew the entire and complete cycle of life, ultimately including the whole cosmos present in its specific realizations, so that in the last analysis one might describe ceremonials as the cosmos becoming thankfully aware of itself."[17]

Carpenter describes this for the Inuit as the individual belief in humans being "he who releases life inherent in nature and guides its expression into beautiful forms."[18] Martin adds that for the Navajo people, who see themselves as having been created "to restore the beautiful," "man experiences beauty by creating it."[19]

My father-in-law, standing in the middle of the Arrow

River waiting on a trout, might have to think about what he can offer of himself. Knowing Jim, he whistles some songs. He soaks in the beauty of the day and appreciates how it feeds him. He thinks about collecting a bouquet of spring lupines to bring to his wife. He watches the flies dancing above the river and with his own homemade fly on his line tries to mimic their movement. Bit by bit he learns their dance. The flies cease to be a nuisance and instead become a set of good teachers from which he could learn something if he wants to bring home a trout. What does he ultimately offer of himself toward creating the beauty of the world? His song, his humor, an open heart, and the humility of a student of flies. The nature writer Barry Lopez would no doubt remind an anthropomorphizer like my father-in-law that "A person cannot be afraid of being foolish. For everything, every gesture, is sacred."[20]

I would like to return to my mother-in-law for a moment, driving along the road in Pelorus Sound, or my father-in-law knee deep in the Arrow River with nary a trout in sight. How are they participating in a gifting relationship with the world around them? They do not need to hunt to acquire food (thank goodness we've never had to rely on Dad's fishing prowess to save us) nor do they need to offer up their skills at writing or watchmaking. They were not raised to sing songs to the sunrise nor have they ever been taught to believe that they are the "artist-regenerators" of the world.

Why, then, do they anthropomorphize their world?

When I try to answer this for my mother-in-law, who has lived away from the Sounds for nearly seventy years of her life, it is that she is ever thankful for the gift of community she once felt there. At a time in her life when her parents had no emotional energy for her, the very alive world around her did. It entered into her imagination and has lived there, as a loved and safe space, ever since. She has never felt truly alone, even when she was abandoned by her parents at The Pines, because the land itself never deserted her. She has painted the Sounds many times and written books about her childhood there, and my father-in-law tells us that she dreams herself there. Surely she could keep all of these things for herself as commodities, like pictures in her scrapbook, but she believes that she owes the place, and all of its beings, a level of gratitude that goes well beyond herself. She believes she is forever in its debt for its generous kindness to her.

When Pam had the opportunity to revisit these places, which have nourished her creativity, nourished her sense of self, and taught her about the capacity to love, she has offered up the greatest gift she has—her unabashed love—and her willingness to greet each bay, each corner, each familiar view, as the old friends they are.

"*Hello Te Towaka,*" she calls coming down over a steep saddle into familiar bush. "*Now how are you?*"

The hills answer her with the mere statement that they are still there to greet her. Generations of trees and birds have lived and died, but the sun still hits the top of Mount Shewell just as she remembers it. Life in all its myriad, confusing, and wonderful ways has continued on. Nature has conserved her not she it. I imagine she will not mind becoming a part of that land when she dies as she has always been a part of it in her life.

In an anthropocentric world, there is no language for the love relationship many feel with the world around them. There are no ceremonies, no ways of expressing the belief that there is hopefulness and possibility to be found in the continuance of life as a whole. There are no celebrations of the foolish gesture. But just because that language does not exist, and because that relationship is not explicitly and culturally valued, does not mean that the lived practice cannot be a deep part of who we are.

In return for offering up a generous heart, respectfulness, and a promise of promoting the future of the collective over the fears of the individual, my in-laws and so many others have received unexpected gifts. There is no eternal life for any one of us, and death will certainly arrive right on time, but there must be great joy to experience death as the returning of a gift and being able, with great generosity, to offer truly all of ourselves to the creative future of the world. For what greater gift can we offer not only to be at home in the world but to become another's home, too?

Epilogue
Inheritance

> Though I do not believe that a plant will spring up
> where no seed has been, I have great faith in a seed.
> Convince me that you have a seed there and I am pre-
> pared to expect wonders.
>
> HENRY DAVID THOREAU, *Faith in a Seed:*
> *The Dispersion of Seeds and Other Late Natural Writings*

Once upon a time, the word *inherit* did not mean "A coming into, or taking possession of something as one's birthright, possession, ownership, right of possession" but instead meant the choice of choosing one's heirs and what one would leave to them, "to make heir, put in possession, cause to inherit."[1] Here, in the notion of inheritance, is the root of the possibility for change. Each of us can choose what we want to pass on, to whom, and how. What will we decide our wealth to be? What wealth will we pass on to them? Will their wealth be found in the landfills of New Jersey, in the song of blackbirds singing outside our doors, or in our memories and stories of each other?

The act of remembering is the antithesis of dismembering as we literally put back together that which is fractured.[2] Perhaps we can remember, we can replace the memories, and we can offer our heirs the comfort of knowing that they have been unambiguously loved by us. The autumn leaf does not fear death; it is most glorious in the moment when it looses itself from the tree and dances to the ground. This is something it only does once, and it has to die to do so, but in its death and decay come all of the possibilities of a beautiful future. Is there a better way to nourish heirs than to become the very place from which they emerge, a home for their new roots?

In the summer we were married, Kevin and I planted a tree at the edge of the cemetery where we, his parents, and sister will eventually be buried. His young nephew is already buried there, a life stopped short by a cancer that entered and took root in his brain. Our tree, a small kowhai, a native tree of New Zealand and one beloved by the liquid-voiced tui birds, is nestled in among totara and pines. Pam and Jim go and tend it when we are away. They write us long letters and tell us stories of "Sophia the Kowhai"—how she weathered the drought, how she avoided the mowers, how Jim has cut the grass at her base to make sure she gets plenty of rainwater, and how she has grown. Always how she has grown.

Someday, when we are again planting and planted in that field, I hope Sophia the Kowhai will have branches strong enough to support the tuis who will come to sing and maybe offer some shade to those who might want to come to sit and watch the wind playing in the tall grass in the meadow above. She might offer a bit of solace to those who will need to remember that Sophia was tended by people who will never see her as full grown, and yet they tended her and loved her as a member of the family, which she is. One of all of our relations.

In a field in New Jersey, one swept by the wings of redtailed hawks, my parents lie in the ground. Three springs ago we started noticing perfectly round holes appearing over my father's grave. Four, five, six holes at regular intervals penetrating the mounded soil. We wondered if the caretakers had some tool and were poking him like one would with a meat thermometer to see if he was truly "done."

Over the seasons we returned as the mystery continued to challenge us. Each time, the holes looked as fresh around the edges as they had the last time we'd come. Surely the caretakers had better things to do with their time than poke holes in dead people!

One morning, as we finished planting black-eyed Susans, we discovered the answer to our riddle. A meadow vole popped its head out of one of the holes and scurried across to another behind the stone. At last we understood the hawks swooping low and understood that once planted in the ground my father had truly returned his body to the community. This was his ultimate gift of life. As someone who had studied mammals, and small mammals in particular, he would have loved knowing that he had become a habitat himself.

Somewhere in the planting and tending of Sophia, in the wonder of a family of voles and the sweep of red-tailed hawks, in the telling of stories, in the giving of life, and in the knowledge that each gift is a seed that may or may not blossom, comes a deepening of faith in the possibility that my human kin will remember that we have always lived in and loved our world. If we can find our way to articulate and honor that love, we may be able to find Thoreau's faith in a seed and a future for us all.

Notes

CHAPTER 1

1. Keith Basso, *Wisdom Sits in Places: Landscape and Language among the Western Apache* (Albuquerque: University of New Mexico Press, 1996), 106.

2. Richard Van Gelder to George Stofysky, August 16, 1976.

3. Terry Tempest Williams, *Pieces of White Shell: A Journey to Navajoland* (New York: Scribners, 1984), 135.

4. Terry Tempest Williams, "Dialogue Two: Landscape, People and Place Robert Finch and Terry Tempest Williams," in *Writing Natural History: Dialogues with Authors,* ed. Ed Lueders, 46 (Salt Lake City: University of Utah Press, 1989).

CHAPTER 2

1. Barry Lopez, *About This Life* (New York: Random House, 1999), 206.

2. Many thanks to Jody Marquis for pointing me toward the term *wildland* and for elucidating the distinction between a wilderness and a wildland.

3. Public Law 88–577, 88th Cong., S. 4, 3 September 1964.

4. Calvin Luther Martin, *In the Spirit of the Earth: Rethinking History and Time* (Baltimore: Johns Hopkins University Press, 1992), 45.

5. David Abram, *The Spell of the Sensuous* (New York: Vintage, 1996), 101.

6. Jack Turner, *The Abstract Wild* (Tucson: University of Arizona Press, 1995), 25.

7. Joseph Meeker, "Comedy and a Play Ethic," http://www.the greatcosmicjoke.com/reallife/playethic.html (June 6, 2003).

8. Linda Hogan, quoted in Derrick Jensen, *Listening to the Land: Conversations about Nature, Culture, and Eros* (San Francisco: Sierra Club, 1995), 123.

9. Gary Snyder, *The Practice of the Wild* (New York: North Point Press, 1990), 23.

CHAPTER 3

1. Wallace Stegner, "Coda: Wilderness Letter," in *The Norton Anthology of Nature Writing*, ed. John Elder and Robert Finch (New York: Norton, 1990), 569.

2. John Muir, *My First Summer in the Sierra* (San Francisco: Sierra Club, 2002).

3. Edmund Carpenter, *Eskimo Realities* (New York: Henry Holt, 1973), 43.

4. Paul Gruchow, *Grassroots: The Universe of Home* (Minneapolis: Milkweed Editions, 1995), 130.

5. Leonard Shlain. *Art and Physics: Parallel Visions in Space, Time, and Light* (New York: Quill William Morrow,1991), 164.

6. Henry David Thoreau, "Walking," in *The Portable Thoreau*, ed. Carl Bode (New York: Penguin, 1982), 593.

7. Patricia Monaghan, interview, February 9, 2001, www.ksharpe.com/road_scholars.

8. Robert Penn Warren, *New and Selected Poems, 1923–1985* (New York: Random House, 1984), 60.

CHAPTER 4

1. Simon Coleman and John Elsner, *Pilgrimage Past and Present: Sacred Travel and Sacred Space in the World Religions* (London: British Museum Press, 1995), 6.

2. Keith Basso, *Wisdom Sits in Places: Landscape and Language among the Western Apache* (Albuquerque: University of New Mexico Press, 1996), 7.

3. Paul Gruchow, *Grassroots: The Universe of Home* (Minneapolis: Milkweed Editions, 1995), 4.

4. Gaston Bachelard, *The Poetics of Space* (Boston: Beacon, 1994), 15.

5. Edith Cobb, *The Ecology of Imagination in Childhood* (New York: Columbia University Press, 1977), 27.

6. Bachelard, *The Poetics of Space*, 6.

7. Gruchow, *Grassroots*, 7.

CHAPTER 5

1. Many thanks to Sally Palmer Thomason for reminding me of this important point.

2. Henry David Thoreau, *Walden* (Boston: Shambhala, 1992), 8.

3. Gaston Bachelard, *The Poetics of Space* (Boston: Beacon, 1994), 6.

4. Roger Hart, *Children's Experience of Place* (New York: Irvington Press, 1979), 285.

5. Sophie Watson, "A Home Is Where the Heart Is: Engendering Notions of Homelessness," in *Homelessness: Exploring the New Terrain,* ed. Patricia Kennett and Alex Marsh (Bristol: Policy Press, University of Bristol, 1999), 85.

6. Daniel Nettle and Suzanne Romaine, *Vanishing Voices: The Extinction of the World's Languages* (Oxford: Oxford University Press, 2000), 14.

7. Leslie Marmon Silko, *Yellow Woman and the Beauty of the Spirit: Essays on Native American Life Today* (New York: Simon and Schuster, 1996), 50.

8. Keith Basso, *Wisdom Sits in Places: Landscape and Language among the Western Apache* (Albuquerque: University of New Mexico Press, 1996), 103.

CHAPTER 6

1. Percy Bysshe Shelley, "Ozymandias," in *Poems* (London: Penguin, 1956), 107.

2. Georg Simmel, "The Ruin," in *Simmel on Culture: Selected Writings,* ed. David Frisby and Mike Featherstone (London: Sage, 1997), 259.

3. David Lewis Williams, *The Mind in the Cave* (London: Thames and Hudson, 2002), 11.

4. Andrew Schelling, *Wild Form and Savage Grammar* (Albuquerque: La Alameda, 2003), 12.

5. Ibid., 13.

6. Ibid.

7. Jack Turner, *The Abstract Wild* (Tucson: University of Arizona Press, 1995), 11.

8. Ibid., 18.

9. Gary Snyder, *The Practice of the Wild* (New York: North Point, 1990), 89.

10. Schelling, *Wild Form and Savage Grammar,* 12.

11. Terry Tempest Williams, *Leap* (New York: Pantheon, 2000), 118.

12. Brian Sykes, *The Seven Daughters of Eve* (London: Corgi, 2001), 15.

13. Snyder, *The Practice of the Wild,* 94.

CHAPTER 7

1. For an excellent description of the neurobiology of this phenomenon, see V. S. Ramachandran and Sandra Blakeslee, *Phantoms in the Brain: Human Nature and the Architecture of the Mind* (London: Fourth Estate, 1998).

2. Vine Deloria, *Red Earth, White Lies: Native Americans and the Myth of Scientific Fact* (New York: Scribners, 1995), 55.

3. Calvin Luther Martin, *In the Spirit of the Earth: Rethinking History and Time* (Baltimore: Johns Hopkins University Press, 1992), 15.

4. Ibid., 15.

5. Stewart Guthrie, *Faces in the Clouds: A New Theory of Religion* (Oxford: Oxford University Press, 1993), 112.

6. Ibid., 5.

7. Martin, *In the Spirit of the Earth,* 20.

8. Edmund Carpenter, *Eskimo Realities* (New York: Holt, Rinehart and Winston, 1973), 106.

9. Leslie Marmon Silko, "Landscape, History, and the Pueblo Imagination," in *The Norton Anthology of Nature Writing,* ed. John Elder and Robert Finch (New York: Norton 1990), 893.

10. Paul Shepard, *The Others: How Animals Made Us Human* (Washington, DC: Island, 1997), 36.

11. Martin, *In the Spirit of the Earth,* 47.

12. Vine Deloria, *God Is Red: A Native View of Religion* (Golden, CO: Fulcrum, 1994), 88.

13. Ibid., 183.

14. Lewis Hyde, *The Gift: Imagination and the Erotic Life of Property* (New York: Vintage, 1983), 4.

15. Calvin Luther Martin, *The Way of the Human Being* (New Haven: Yale University Press, 1999), 40.

16. "Diane Burns," in Joseph Bruchac, *Survival This Way: Interviews with American Indian Poets* (Tucson: Sun Tracks and University of Arizona Press, 1987), 55.

17. Deloria, *God Is Red,* 130.

18. Carpenter, *Eskimo Realities*, 42.

19. Martin, *The Way of the Human Being*, 24.

20. Barry Lopez, *River Notes: The Dance of Herons* (New York: Avon, 1979), 80.

EPILOGUE

1. Definition from the *Oxford English Dictionary*.

2. Daniel Taylor, *Tell Me a Story: The Life Changing Power of Stories* (Saint Paul: Bog Walk, 2001), 38.

Bibliography

Abram, David. 1996. *The spell of the sensuous.* New York: Vintage.

Auge, Marc. 1995. *Non-places: Introduction to an anthropology of super-modernity.* London: Verso.

Awiakta, Marilou. 1993. *Selu: Seeking the corn mother's wisdom.* Golden, CO: Fulcrum.

Bachelard, Gaston. 1994. *The poetics of space.* Boston: Beacon.

Basso, Keith. 1996. *Wisdom sits in places: Landscape and language among the western Apache.* Albuquerque: University of New Mexico Press.

"Diane Burns." 1987. In Joseph Bruchac, *Survival This Way: Interviews with American Indian Poets.* Tucson: Sun Tracks and University of Arizona Press.

Callicott, J. Baird, and Michael P. Nelson, eds. 1998. *The great new wilderness debate.* Athens: University of Georgia Press.

Carpenter. Edmund. 1973. *Eskimo realities.* New York: Holt, Rinehart and Winston.

Casey, Edward S. 1997. *The fate of place: A philosophical history.* Berkeley: University of California Press.

Chamberlain, J. Edward. 2004. *If this is your land, where are your stories? Finding common ground.* Toronto: Vintage Canada.

Cobb, Edith. 1977. *The ecology of imagination in childhood.* New York: Columbia University Press.

Coleman, Simon, and John Elsner. 1995. *Pilgrimage past and present: Sacred travel and sacred space in the world religions.* London: British Museum Press.

Coles, Robert. 1989. *The call of stories: Teaching and the moral imagination.* Boston: Houghton Mifflin.

Cresswell, Tim. 2004. *Place: A short introduction.* Oxford: Blackwell.

De Botton, Alain. 2006. *The architecture of happiness.* London: Penguin.

Deloria, Vine. 1994. *God is red: A native view of religion.* Golden, CO: Fulcrum.

Deloria, Vine. 1995. *Red earth, white lies: Native Americans and the myth of scientific fact*. New York: Scribners.

Dunbar, Robin. 1996. *Grooming, gossip, and the evolution of language*. London: Faber and Faber.

Ehrlich, Gretel. 1985. *The solace of open spaces*. New York: Viking.

Elder, John, and Robert Finch, eds. 1990. *The Norton anthology of nature writing*. New York: Norton.

Engle, J. Ronald. 1992. "Renewing the bond of mankind and nature: Biosphere reserves as sacred space." In *Finding home*, ed. Peter Sauer. Boston: Beacon.

Feld, Steven, and Keith Basso, eds. 1997. *Senses of place*. New York: School of American Research Press.

Gallagher, Winifred. 1993. *The power of place: How our surroundings shape our thoughts, emotions, and actions*. New York: Poseidon.

Glacken, Clarence J. 1967. *Traces on the Rhodian shore: Nature and culture in western thought from ancient times to the end of the eighteenth century*. Berkeley: University of California.

Gruchow, Paul. 1988. *The necessity of empty places*. New York: St. Martin's.

Gruchow, Paul. 1995. *Grassroots: The universe of home*. Minneapolis: Milkweed Editions.

Gruchow, Paul. 1997. *Boundary waters: The grace of the wild*. Minneapolis: Milkweed Editions.

Guthrie, Stewart. 1993. *Faces in the clouds: A new theory of religion*. Oxford: Oxford University Press.

Hart, Roger. 1979. *Children's experience of place*. New York: Irvington.

Hayden, Dolores. 1995. *The power of place: Urban landscapes as public history*. Cambridge: MIT Press.

Hiss, Tony. 1990. *The experience of place*. New York: Knopf.

Hogan, Linda. 1995. *Dwellings: A spiritual history of the living world*. New York: Touchstone.

Hyde, Lewis. 1983. *The gift: Imagination and the erotic life of property*. New York: Vintage.

Jacobs, Jane. 1961. *The death and life of great American cities*. New York: Vintage.

Jensen, Derrick. 1995. *Listening to the land: Conversations about nature, culture, and eros*. San Francisco: Sierra Club.

Kazin, Alfred. 1988. *A writer's America: Landscape in literature*. New York: Knopf.

Kemmis, Daniel. 1990. *Community and the politics of place*. Norman: University of Oklahoma Press.

Kittredge, William. 2000. *The nature of generosity*. New York: Vintage.

Krakauer, Jon. 1996. *Into the wild*. New York: Villard.

Kuntsler, James. 1994. *The geography of nowhere.* New York: Simon and Schuster.

Langer, Susanne. 1953. *Feeling and form: A theory of art.* New York: Charles Scribner and Sons.

Lefebvre, Henri. 2000. *The production of space.* Oxford: Blackwell.

Lewis-Williams, David. 2002. *The mind in the cave.* London: Thames and Hudson.

Lippard, Lucy R. 1995. *The lure of the local: Senses of place in a multi-centered society.* New York: New Press.

Lippard, Lucy R. 1999. *On the beaten track: Tourism, art, and place.* New York: New Press.

Lopez, Barry. 1979. *River notes: The dance of herons.* New York: Avon.

Lopez, Barry. 1988. *Crossing open ground.* New York: Charles Scribner and Sons.

Lopez, Barry. 1999. *About this life.* New York: Random House.

Lueders, Ed. 1989. *Writing natural history: Dialogues with authors.* Salt Lake City: University of Utah Press.

Macaulay, David. 1979. *Motel of the mysteries.* Boston: Houghton Mifflin.

Macy, Joanna. 1991. *World as lover, world as self.* Berkeley: Parallax.

Mander, Jerry. 1991. *In the absence of the sacred: The failure of technology and the survival of the Indian nations.* San Francisco: Sierra Club.

Marshall III, James. 1995. *On behalf of wolf and the first peoples.* Santa Fe: Red Crane.

Marshal III, James. 1998. *The dance house.* Santa Fe: Red Crane.

Marshall, Peter. 1994. *Nature's web: Rethinking our place on earth.* New York: Paragon House.

Martin, Calvin Luther. 1992. *In the spirit of the earth: Rethinking history and time.* Baltimore: Johns Hopkins University Press.

Martin, Calvin Luther. 1999. *The way of the human being.* New Haven: Yale University Press.

Marx, Leo. 1964. *The machine in the garden: Technology and the pastoral ideal in America.* New York: Oxford University Press.

Matthews, M. H. 1992. *Making sense of place: Children's understanding of large-scale environments.* New York: Harvester Wheatsheaf.

Matthiessen, Peter. 1978. *The snow leopard.* New York: Viking.

Meeker, Joseph W. 1997. *The comedy of survival: Literary ecology and a play ethic.* Tucson: University of Arizona Press.

Mithen, Steven. 1996. *The prehistory of the mind: A search for the origins of art, religion, and science.* London: Thames and Hudson.

Momaday, N. Scott. 1976. *The names.* Tucson: University of Arizona Press.

Momaday, N. Scott. 1997. *The man made of words.* New York: St. Martin's.

Monaghan, Patricia. 2003. *The red haired girl from the bog*. Novato, CA: New World Library.

Muir, John. 2002. *My first summer in the Sierra*. San Francisco: Sierra Club.

Myerhoff, Barbara, et al. 1992. *Remembered lives*. Ann Arbor: University of Michigan Press.

Nabhan, Gary Paul. 1997. *Cultures of habitat: On nature, culture, and story*. Washington, DC: Counterpoint.

Nabhan, Gary Paul, and Stephen Trimble. 1994. *The geography of childhood: Why children need wild places*. Boston: Beacon.

Nabokov, Peter. 2002. *A forest of time: American Indian ways of history*. Cambridge: Cambridge University Press.

Nash, Roderick. 1967. *Wilderness and the American mind*. New Haven: Yale University Press.

Nelson, Richard K. 1983. *Make prayers to raven: A Koykon view of the northern forest*. Chicago: University of Chicago Press.

Nelson, Richard K. 1989. *The island within*. New York: Vintage.

Nettle, Daniel, and Suzanne Romaine. 2000. *Vanishing voices: The extinction of the world's languages*. Oxford: Oxford University Press.

Niles, John N. 1991. *Homo narrans: The poetics and anthropology of oral literature*. Philadelphia: University of Pennsylvania Press.

Oelschlager, Max. 1991. *The idea of wilderness: From prehistory to the age of ecology*. New Haven: Yale University Press.

Peat, F. David. 1994. *Blackfoot physics: A journey into the Native American universe*. London: Fourth Estate.

Ramachandran, V. S., and Sandra Blakeslee. 1998. *Phantoms in the brain: Human nature and the architecture of the mind*. London: Fourth Estate.

Robinson, Tim. 1996. *Setting foot on the shores of Connemara and other writings*. Dublin: Lilliput.

Robinson, Tim. 2001. *My time in space*. Dublin: Lilliput.

Robinson, Tim. 2006. *Connemara: Listening to the wind*. London: Penguin.

Rodaway, Paul. 1994. *Sensuous geographies: Body, sense, and place*. London: Routledge.

Roszak, Theodore. 1992. *The voice of the earth*. New York: Simon and Schuster.

Rubin, Daniel. 1995. *Memory in oral traditions: The cognitive psychology of epic, ballads, and counting-out rhymes*. Oxford: Oxford University Press.

Ryden, Kent C. 1993. *Mapping the invisible landscape: Folklore, writing, and the sense of place*. Iowa City: University of Iowa Press.

Sack, Robert David. 1992. *Place, modernity, and the consumer's world: A regional framework for geographical analysis.* Baltimore: Johns Hopkins University Press.

Sanders, Scott Russell. 1991. *Secrets of the universe: Scenes from the journey home.* Boston: Beacon.

Schelling, Andrew. 2003. *Wild form and savage grammar.* Albuquerque: La Alameda.

Schumacher, E. F. 1973. *Small is beautiful: Economics as if people mattered.* New York: Harper and Row.

Seamon, David, and Robert Mugerauer, eds. 1985. *Dwelling, place, and environment: Towards a phenomenology of person and world.* New York: Columbia University Press.

Shelley, Percy Bysshe. 1956. "Ozymandias." In *Poems.* London: Penguin.

Shepard, Paul. 1995. *The only world we've got.* Washington, DC: Island.

Shepard, Paul. 1997. *The others: How animals made us human.* Washington, DC: Island.

Shlain, Leonard. 1991. *Art and physics: Parallel visions in space, time, and light.* New York: Quill William Morrow.

Silko, Leslie Marmon. 1996. *Yellow woman and a beauty of the spirit: Essays on Native American life today.* New York: Simon and Schuster.

Simmel, Georg. 1997. "The ruin." In *Simmel on Culture: Selected Writings,* ed. David Frisby and Mike Featherstone. London: Sage.

Slovic, Scott, and Terrell Dixon. 1993. *Being in the world: An environmental reader for writers.* New York: Macmillan.

Smith, Henry Nash. 1950. *Virgin land: The American West as symbol and myth.* Cambridge: Harvard University Press.

Snyder, Gary. 1980. *The real work: Interviews and talks, 1964–1979.* New York: New Directions.

Snyder, Gary. 1990. *The practice of the wild.* New York: North Point.

Snyder, Gary. 1995. *A place in space: Ethics, aesthetics, and watersheds.* Washington, DC: Counterpoint.

Stegner, Wallace. 1990. "Coda: Wilderness letter." In *The Norton Anthology of Nature Writing,* ed. John Elder and Robert Finch. New York: Norton.

Stegner, Wallace. 1992. *Where the bluebird sings to the lemonade springs: Living and writing in the West.* New York: Wings.

Swimme, Brian, and Thomas Berry. 1992. *The universe story.* San Francisco: Harper Collins.

Sykes, Brian. 2001. *The seven daughters of Eve.* London: Corgi.

Tallmadge, John. 1997. *Meeting the tree of life.* Salt Lake City: University of Utah Press.

Tattersall, Ian. 1998. *Becoming human: Evolution and human uniqueness.* London: Harcourt Brace.

Taylor, Daniel. 2001. *Tell me a story: The life changing power of stories.* Saint Paul: Bog Walk.

Thomason, Sally Palmer. 2006. *The living spirit of the crone: Turning aging inside out.* Minneapolis: Augsburg.

Thoreau, Henry David. 1982. "Walking." In *The portable Thoreau,* ed. Carl Bode. New York: Penguin.

Thoreau, Henry David. 1992. *Walden; or life in the woods.* Boston: Shambhala.

Thoreau, Henry David. 1993. *Faith in a seed.* Washington, DC: Island.

Trimble, Steve, ed. 1996. *Testimony: Writers of the West speak on behalf of Utah wilderness.* Minneapolis: Milkweed Editions.

Tuan, Yi Fu. 1974. *Topophilia: A study of perception, attitudes, and values.* Englewood Cliffs, NJ: Prentice-Hall.

Tuan, Yi Fu. 1995. *Passing strange and wonderful: Aesthetics, nature, and culture.* New York: Kodansha International.

Tuan, Yi Fu. 2001. *Space and place: The perspective of experience.* Minneapolis: University of Minnesota Press.

Turner, Frederick. 1989. *The spirit of place: The making of an American literary landscape.* San Francisco: Sierra Club.

Turner, Jack. 1995. *The abstract wild.* Tucson: University of Arizona Press.

Vitek, William, and Wes Jackson, eds. 1996. *Rooted in the land: Essays on community and place.* New Haven: Yale University Press.

Warren, Robert Penn. 1984. *New and selected poems, 1923–1985.* New York: Random House.

Watson, Sophie. 1999. "A home is where the heart is: Engendering notions of homelessness." In *Homelessness: Exploring the new terrain,* ed. Patricia Kennett and Alex Marsh. Bristol: Policy Press, University of Bristol.

White, Jonathan. 1994. *Talking on the water: Conversations about nature and creativity.* San Francisco: Sierra Club.

Williams, Terry Tempest. 1984. *Pieces of white shell: A journey to Navajoland.* New York: Scribner.

Williams, Terry Tempest. "Landscape, People, and Place." In *Writing Natural History: Dialogues with Authors,* ed. Ed Lueders. Salt Lake City: University of Utah Press.

Williams, Terry Tempest. 1991. *Refuge: An unnatural history of family and place.* New York: Vintage.

Williams, Terry Tempest. 1994. *An unspoken hunger: Stories from the field.* New York: Pantheon.

Williams, Terry Tempest. 2000. *Leap.* New York: Pantheon.

Wilson, Alexander. 1992. *The culture of nature*. Cambridge, MA: Blackwell.

Wilson, Edward O. 1998. *Consilience: The unity of knowledge*. New York: Knopf.

Wittgenstein, Ludwig. 2001. *Philosophical investigations*. Trans. G. E. M. Anscombe. 3d ed. Oxford: Blackwell.

Woodard, Charles. 1989. *Ancestral voice: Conversations with N. Scott Momaday*. Lincoln: University of Nebraska Press.

Zipes, Jack. 1995. *Creative storytelling: Building community, changing lives*. New York: Routledge.

Index

Text design by Mary H. Sexton
Typesetting by Delmastype, Ann Arbor, Michigan
Font: New Baskerville

"British printer John Baskerville of Birmingham created the types that bear his name in about 1752. George Jones designed this version of Baskerville for Linotype-Hell in 1930, and the International Typeface Corporation licensed it in 1982."
—Courtesy www.adobe.com